BRITISH GEOLOGICAL SURVEY

R A EDWARDS

# The Minehead district — a concise account of the geology

Memoir for 1:50 000 Geological Sheet 278 and part of sheet 294 (England and Wales)

CONTRIBUTORS

*Offshore geology*
C D R Evans
D H Jeffery

*Geophysics*
J D Cornwell
C P Royles

*Seismic interpretation*
S Holloway
N J P Smith

*Stratigraphy*
A Whittaker

*Biostratigraphy*
B M Cox
H C Ivimey-Cook
G Warrington
J B Riding
I P Wilkinson
N J Riley

*Sedimentology*
N S Jones

*Hydrogeology*
R J Marks

*Petrology*
G E Strong

*Engineering geology*
A Forster

*Mineralisation*
R C Scrivener

*Radon*
T K Ball

*Remote sensing*
S H Marsh

London: The Stationery Office    1999

The grid used on the figures is the National Grid taken from the Ordnance Survey maps. Figure 2 is based on Ordnance Survey 1:50 000 scale maps, numbers 180 and 181.
© Crown copyright reserved.
Ordnance Survey licence no. GD272191/1999.

ISBN 0 11 884 544 6

*Bibliographical reference*

EDWARDS, R A. 1999.    The Minehead district — a concise account of the geology.    *Memoir of the British Geological Survey*, Sheet 278 and part of Sheet 294 (England and Wales).

AUTHOR

R A Edwards, BSc, PhD
*British Geological Survey, Exeter*

CONTRIBUTORS

C D R Evans, BSc, PhD
C P Royles, BSc
S Holloway, BA, PhD, CGeol
N J P Smith, MSc
A Whittaker, BSc, PhD, CGeol
B M Cox, BSc, PhD, CGeol
G Warrington, DSc, CGeol
J B Riding, BSc, PhD, CGeol
I P Wilkinson, BSc, PhD, CGeol
N J Riley, BSc, PhD, CGeol
N S Jones, BSc, PhD
A Forster, BSc, CGeol
S H Marsh, BSc, PhD
G E Strong, BSc
*British Geological Survey, Keyworth*

R J Marks, BSc, CGeol
*British Geological Survey, Wallingford*

R C Scrivener, BSc, PhD, CGeol
*British Geological Survey, Exeter*

T K Ball, BSc, PhD
D H Jeffery, BSc, CGeol, Eur Geol
J D Cornwell, MSc, PhD
H C Ivimey-Cook, BSc, PhD
formerly *British Geological Survey, Keyworth*

Printed in the UK for the Stationery Office

J100946   C6   12/99

## Acknowledgements

This memoir has been written and compiled by Dr R A Edwards, and incorporates the specialist contributions listed below.

The account of the offshore geology is based on a report by Dr C D R Evans and Mr D H Jeffery. Dr S Holloway and Mr N J P Smith have reported on their interpretation of deep offshore seismic profiles. The seismic sections illustrated and described come from a variety of non-exclusive proprietary surveys owned by Geco-Prakla, a division of Geco Geophysical Company Limited. Permission to use these data is gratefully acknowledged.

Dr A Whittaker has contributed the results of his detailed mapping of the Blue Anchor-Watchet foreshore, carried out using 1:5000 scale aerial photographs as an extension of the survey of the Weston-super-Mare district (Sheet 279) in 1970 and 1975; he has also provided data from his study of the Selworthy boreholes.

Dr H C Ivimey-Cook has provided accounts of the Penarth Group and the Lias Group of the district. He has reported on the sequence and biostratigraphy of the Jurassic rocks in the two cored boreholes (72/63; 73/62) in the offshore area of the district, and has also described the sequence and macrofossils (particularly ammonites) of the Selworthy boreholes. Dr G Warrington has provided an account of the palynology of the Late Triassic to earliest Jurassic strata of the Selworthy No. 2 Borehole, and has reported on the palynology of samples collected by Dr Edwards from outcrops of Mercia Mudstone in the district. Dr B M Cox has commented on the stratigraphy of the offshore Jurassic sequence, and has provided a list of samples with biostratigraphical attributions. Dr N J Riley commented on the identification of Carboniferous Limestone clasts in thin sections of the Luccombe Breccia and Budleigh Salterton Pebble Beds. The palynology and calcareous microfaunas of four gravity core samples from the offshore Jurassic outcrop were reported on by Dr J B Riding and Dr I P Wilkinson, respectively.

Mr and Mrs N Rodber of Minehead kindly donated a specimen of the plant fossil *Pseudosporochnus nodosus* found by them in Henner's Combe, north of Selworthy Beacon, and identified by Professor Dianne Edwards of the University of Wales College of Cardiff.

Dr N S Jones has reported on his investigations of the sedimentology of the Hangman Sandstone and the Permo-Triassic in the district.

Dr J D Cornwell has contributed a report on the geophysics of the district. He and Mr C P Royles carried out a gravity survey and made resistivity soundings in the Porlock Basin.

# Geology of the Minehead district

The land area described in this memoir lies mainly within west Somerset, south-west England. Much of it is scenically attractive and lies within Exmoor National Park; tourism and agriculture are important activities. The offshore part of the district extends into the Bristol Channel. The centres of population are Minehead, Porlock and Watchet.

Three main sedimentary rock divisions, of Devonian, Permo-Triassic to Jurassic, and Quaternary age are present at outcrop. Devonian rocks, mainly sandstones, occupy about 75 per cent of the land area. Red breccias, conglomerates, sandstones and mudstones of the Permo-Triassic occupy partly faulted (half-graben) basins surrounded by hills of Devonian rocks. The Jurassic rocks are mainly grey mudstones with some limestones; they occupy most of the offshore area. Quaternary deposits, especially periglacial head, are widespread. Away from the coast, Quaternary deposits are absent from much of the sea bed.

Precambrian crystalline basement rocks are inferred at depths of over 6 km beneath the Bristol Channel. Offshore seismic profiles suggest that Lower Cambrian quartzites are probably present, but Ordovician and Silurian rocks are not known definitely to occur beneath the district.

The oldest exposed rocks are shallow marine slates and sandstones of the Lynton Formation (late Emsian to early Eifelian), which has a small outcrop. The succeeding Hangman Sandstone Formation (mainly Eifelian) is a thick sequence of sandstones with subordinate mudstones, deposited on a distal alluvial fan and ephemeral continental mudflats.

Carboniferous rocks do not crop out in the district, but reinterpretation of seismic reflection profiles indicates that Carboniferous Limestone (Dinantian) and overlying Silesian rocks may be present in the subcrop beneath much of the offshore area. In the south of the district, the Carboniferous rocks are thought to be overthrust by Devonian rocks.

During the Variscan Orogeny, the Devonian rocks were folded and faulted; the folds trend mainly east-south-east and are mainly overturned to the north. Post-orogenic extension followed formation of the Variscan foldbelt. The New Red Sandstone Supergroup, consisting of material eroded from the Variscan mountains, rests unconformably on folded Devonian rocks. The earliest deposits (Luccombe Breccia) in the onshore Porlock Basin were laid down on the distal part of an alluvial fan flanked by aeolian dunes; a thin conglomerate bed containing rounded boulders was deposited in a fluvial channel. The Luccombe Breccia is probably Triassic, but may be Permian or even latest Carboniferous (Stephanian). It does not occur in the Minehead Basin where possible coeval deposits are the Triassic Budleigh Salterton Pebble Beds and Otter Sandstone.

In Late Triassic times (possibly Carnian to Norian), more uniform conditions were established over the district, and the muds which make up the Mercia Mudstone Group were deposited. Deposition of mixed floodplain and playa lithologies probably took place in a low-relief continental basin. The red Mercia Mudstone (Triassic) to grey Lias Group (Jurassic) succession records the change from a mainly continental environment of playas and supratidal sabkhas, to an epicontinental marine environment. The Rydon Member, forming the bulk of the Blue Anchor Formation at the top of the Mercia Mudstone, contains sulphate nodules and carbonaceous mudstones, of possible algal mat origin, suggesting deposition in a sabkha environment. The onset of a marine transgression from the Tethyan province in the south is first evident in the uppermost beds of this formation. With deposition of the Williton Member, at the top of the Blue Anchor Formation, the environment became fully marine. The Westbury Formation of the Penarth Group was deposited in a transgressive, littoral environment, with lower energy, stagnant or weakly oxygenated water. Contorted beds in the Cotham Member of the overlying Lilstock Formation were probably caused by earthquake activity.

A mainly argillaceous Lower to Upper Jurassic succession was deposited during prolonged subsidence of the Bristol Channel Basin; associated faults (Central Bristol Channel Fault Zone) represent reactivation of a Variscan thrust. The Lower and Middle Lias consist of mudstones with limestone interbeds (Blue Lias Formation) at the base, and were deposited in a shallow epicontinental sea. The Upper Lias consists mainly of marine silty shale and mudstone. Above it, the equivalents of the Inferior Oolite and Great Oolite groups consist of grey marine shales. The Oxford Clay is similar to that onshore in southern England, and was deposited in a shallow sea; it is not separable from the shales beneath. The youngest Jurassic strata in the district are possible Corallian sands, deposited in an estuary or delta.

There is a marked break in the preserved succession between Late Jurassic rocks and the Quaternary deposits. Any Cretaceous, Paleogene and Neogene deposits which may have formed in the district have been removed by erosion.

During the Quaternary glaciations, the area lay south of the ice sheets, but widespread solifluction deposits (head) show that it experienced a periglacial climate. Stages in the Quaternary evolution of the river systems are indicated by river terrace gravels at levels above the present alluvium. Present-day deposition along the river valleys is producing alluvium — mainly silt, clay and gravel. Melting of the last ice sheets resulted in a rise in sea level (the 'Flandrian transgression') which produced marine deposits.

Key issues on which the geology of the district has a bearing are discussed in the memoir. They include slope stability, foundation conditions, artificial deposits, water resources, flood risk, mineral resources, radon, soils and agriculture, and geological conservation.

*Cover photograph*

Blue Anchor Cliff [ST 0385 4368] showing red Mercia Mudstone faulted against the type section of the Blue Anchor Formation, with Penarth Group at the top of the cliff. The top of the Blue Anchor Formation is marked by the change in colour from grey and yellowish brown to black, about three-quarters of the way up the cliff; the base occurs off the photograph to the left. Since this photograph was taken, substantial rock falls have occurred in the vicinity of Blue Anchor Cliff — see Plate 2. (A 11715)
(Photographer: C Jeffrey)

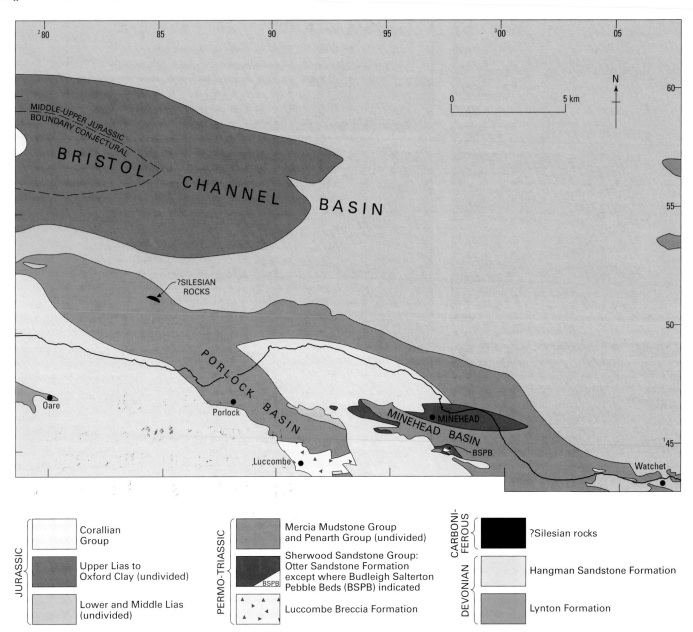

**Figure 1**  Simplified geology of the district. Faults and Quaternary deposits omitted.

Mr G E Strong has provided petrographical descriptions of thin sections from various formations in the district.

Dr S H Marsh provided an interpretation of Landsat TM imagery of the district.

The applied geology chapter includes a section on the engineering geology of the district by Mr A Forster. The account is based on data from the few site investigations available, on information in published papers, and on field observations made during August 1994. Mr R J Marks has contributed an account of the hydrogeology and water supply. The mineralisation of the Luccombe Breccia has been reported on by Dr R C Scrivener.

Dr T K Ball provided data on the radon potential of various formations in the district.

Mr A J Lewis, Director of Development Services, West Somerset District Council, Williton, is thanked for providing borehole records from site investigations within the district.

The assistance of Mr H Prudden of Montacute, Somerset, is gratefully acknowledged. Details of geological localities recorded by him as part of a scheme funded by the Curry Fund are held by the Somerset Environmental Records Centre. Copies were obtained from the Centre and found to be useful in locating and evaluating exposures.

Dr S C Jennings, University of North London, kindly provided details of borehole investigations and radiocarbon dates from the Holocene deposits of the Porlock Bay area.

Mrs H J Wilson of the Field Studies Council at Nettlecombe Court, Williton, is thanked for providing data, incorporated into Chapter 8, on the size and shape of pebbles on the Porlock Bay shingle ridge.

The collaboration and ready assistance of landowners throughout the district is gratefully acknowledged. Exmoor National Park Authority kindly allowed access to aerial photographs of the coastal part of the district. Exmoor National Park Authority and the Countryside Commission kindly gave permission for the use in the memoir and maps of data contained in reports by Integral Geotechnique Ltd on coastal landslips west of Porlock Weir.

Mr M Edgington of the Somerset and Avon Team of English Nature, Taunton, is thanked for providing details of Sites of Special Scientific Interest (SSSIs) within the district.

The memoir was edited by Drs G Warrington and A A Jackson.

## Notes

The word 'district' used in this memoir means the area included in the 1:50 000 Series Sheet 278 (Minehead) and a small part of Sheet 294 (Dulverton).

Figures in square brackets are National Grid References; places within the Minehead district lie within the 100 km squares SS and ST. The grid letters precede the grid numbers.

Dips are given in the form '50° to 287°'. The first number is the angle of dip in degrees and the second is its full-circle bearing, measured clockwise from True North.

Author citations for fossil species mentioned in the text are given on p.122.

Numbers preceded by GS or A refer to photographs in the Geological Survey collections.

Numbers preceded by E refer to thin sections in the National Sliced Rock Collection at the British Geological Survey, Keyworth.

The grain-size scale used in this memoir is the Wentworth (1922) grade scale, modified to the phi ($\emptyset$) scale by Krumbein (1934). The classification of sorting and skewness used is that of Folk and Ward (1957).

Many sites referred to in this memoir are on private land, and permission should be sought from landowners prior to making any visit. The coastal localities are **potentially hazardous** owing to the high tidal range and the long distances between access points.

Maps and diagrams in this memoir use topography based on Ordnance Survey mapping.

# CONTENTS

FIGURES

TABLES

PLATES

# PREFACE

This memoir provides the first modern geological description of the onshore and offshore parts of the Minehead district in west Somerset. The land area, much of which lies within Exmoor National Park, is of considerable scenic variety and beauty; tourism and agriculture are the main economic activities. The many visitors to the area interested in the countryside, as well as professional and amateur geologists attracted to the splendid coastal exposures, will find much of interest in this account. The area has a further significance to me as Director of the Survey in that it is here that my predecessor and founding Director, Henry de la Beche, did his early geological mapping.

The memoir has been arranged so that the information in it is readily accessible to readers requiring a rapid overview of the geology, as well as those needing more detailed information. The first two chapters (One: Introduction, and Two: Applied Geology) are written for the first category of user. The Applied Geology chapter represents a new departure from the traditional style of memoir, and begins with a list of key issues on which geology has a bearing, followed by fuller discussion of those issues. This chapter will be of particular value to planners, engineers and developers. Another departure from previous Survey memoirs is that only selected details of the more important localities are included. Comprehensive details of smaller exposures are given in a separate Technical Report.

The detailed onshore mapping has emphasised the geological and topographical contrast between the upland areas of resistant Devonian sandstones and the low-lying basins of Porlock and Minehead, largely floored by less resistant New Red Sandstone Supergroup (mainly Triassic) strata. These basins are 'half-graben' structures, with large faults along their northern sides. The Porlock Basin sequence extends up into Lower Jurassic (Blue Lias) strata which are the most westerly occurrence of these deposits onshore in England; they have been investigated by cored boreholes. Breccias at the base of the sequence in the Porlock Basin were deposited on an alluvial fan flanked by aeolian dunes. In the Minehead Basin, the equivalent strata may include the fluvial Budleigh Salterton Pebble Beds, which contain pebbles of derived Carboniferous Limestone, and the Otter Sandstone.

From Blue Anchor to Watchet, the cliffs and foreshore show exposures in faulted and folded Late Triassic and Early Jurassic rocks, and include the type section of the Late Triassic Blue Anchor Formation. Offshore, up to 1600 m of largely argillaceous Early to Late Jurassic rocks occupy a major faulted syncline (the Bristol Channel Basin).

Seismic surveys indicate the presence of a northward-dipping shallow reflector, at depths of 2 to 3 km beneath the onshore part of the district, which may be the top of the Carboniferous Limestone. The ExmoorCannington Park Thrust, on which Devonian and older rocks are emplaced over possible Carboniferous Limestone, is believed to cross the district.

Quaternary deposits are widespread, and formed in marine, fluvial and aeolian environments; they also include mass-movement deposits, mainly head which blankets much of the Devonian outcrop.

Landslips occur along much of the coastline. The sea floor has been largely swept clear of superficial sediment by strong tidal currents, but nearer the coast there are accumulations of gravel with some sand. Submarine forest deposits occur in the intertidal zone in Porlock Bay and offshore from Minehead.

Planning considerations in the district are dominated by the fact that much of it lies within a National Park. Many small quarries are scattered over the Devonian sandstone outcrop, but none is currently worked. The sandstones remain a resource of hard rock, but exploitation is unlikely since the outcrops lie mainly within the National Park. The resources of sand and gravel are mainly small. Metalliferous mineralisation is restricted to haematised New Red Sandstone breccias formerly worked near Luccombe. No oil wells have been drilled in the offshore part of the district, and the hydrocarbon potential of the area has generally been considered to be low. However, it is now thought that Silesian strata and possible reservoir rocks (such as Triassic sandstones) may be present there, with significance for hydrocarbon prospectivity. Natural slopes inland are generally stable, but major landslips, some active, occur along much of the coastline. All formations should be adequately investigated for foundation conditions, but particular care is necessary where structures are to be founded on artificial deposits, head, alluvium, saltmarsh deposits, Penarth Group, weathered Mercia Mudstone or weathered Lias. There are no extensive major aquifers in the district, and the main public water supply is derived from springs and boreholes in the Devonian Hangman Sandstone. Some areas near Porlock and Minehead and at the mouth of the Pill River near Blue Anchor are considered to be at risk from tidal flooding.

David A Falvey, PhD
*Director*

*British Geological Survey*
*Kingsley Dunham Centre*
*Keyworth*
*Nottingham*
*NG12 5GG*

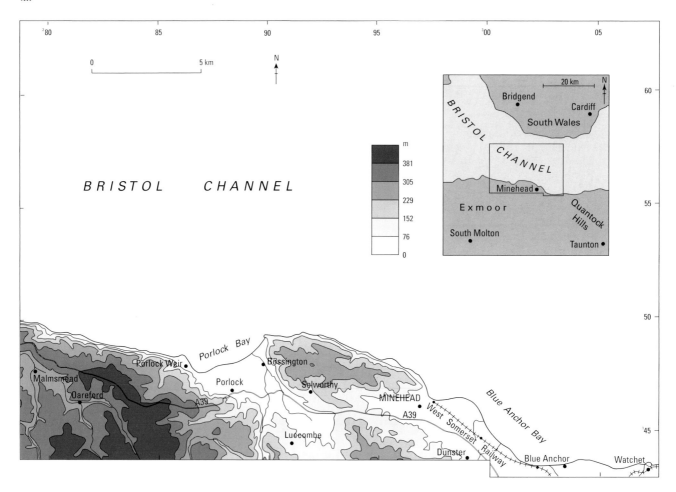

**Figure 2** Topographical map of the district, with main settlements, rivers, roads and railway.

ONE

# Introduction

The area described in this memoir comprises the 1:50 000 Series Sheet 278 (Minehead) and a small part of Sheet 294 (Dulverton). About 100 km$^2$ of the district is land, and lies mainly within west Somerset, except for a small area along the western boundary which forms part of Devon. East of the National Grid Easting $^3$00, near Dunster, the southern boundary of the district has been extended into the Dulverton district as far south as the National Grid Northing $^1$43, in order to include the classical coastal sections between Blue Anchor and Watchet. Offshore, part of the district extends into the Bristol Channel and its geology is also described in the memoir.

The land area contains a variety of attractive coastal and inland scenery and much of it is included within Exmoor National Park. The largest town is the seaside resort of Minehead; other main settlements are Watchet and Porlock. No major rivers are present in the district; Porlock Vale is drained by the Horner Water and its tributaries, while in the east the main streams are the River Avill near Dunster, the Pill River near Blue Anchor and the Washford River near Watchet.

The geological succession in the district is shown in Table 1, and a summary geological map is shown on Figure 1 (p.ii). The oldest outcropping rocks in the district, of Devonian age, extend over about three-quarters of the land area and consist mainly of folded and fractured sandstones. They contrast with the younger Permo-Triassic (New Red Sandstone Supergroup) and Jurassic rocks. The New Red Sandstone consists mainly of red breccias, conglomerates, sandstones and mudstones, and occupies partly faulted onshore basins surrounded by Devonian rocks; it is also present in the subcrop beneath most of the offshore area. The Jurassic rocks, which form most of the sea bed and some small onshore outcrops, are predominantly grey mudstones, with some limestones in the lower part. Quaternary deposits are widespread onshore, particularly on the Hangman Sandstone outcrop, but offshore they have largely been removed from the sea bed by strong tidal currents.

Concealed formations may include Carboniferous Limestone and Silesian rocks beneath much of the offshore area. In the south of the district, Carboniferous rocks are thought to be overthrust by Devonian rocks along a westward continuation of the Exmoor–Cannington Park Thrust. Seismic data indicate the probable presence of Lower Cambrian rocks at depth. Precambrian crystalline rocks are inferred at depths of 5.5 to 6.5 km beneath the district. An outline account of the geology and geological history of the district is given below.

The key geological factors relevant to land-use planning and development in the district are listed and discussed in Chapter 2. Tourism and agriculture are the most important occupations in the district; a light industrial area is located east of Minehead. In the substantial part of the district which lies within Exmoor National Park, the overriding planning policies are concerned with the preservation and enhancement of the natural beauty of Exmoor. Outside the park, the remainder of the district lies within a Special Landscape Area (p.5), and planning policies are directed towards conservation of the character of the landscape. Most of the formations of the district are not likely to give rise to special geotechnical problems, but the Mercia Mudstone, Penarth Group, Blue Lias, marine Quaternary deposits, alluvium and artificial deposits may present foundation problems in certain circumstances (see Chapter 2). In general, slope stability does not appear to be a major problem in the inland areas, but major areas of instability are present on the coastal slopes, and some of the landslips are currently active (Chapter 2).

There is no extractive industry in the area, but many small quarries and pits scattered over the district testify to the former exploitation of most of the formations for varied purposes, including the extraction of building stone, roadstone, brick clay, lime and gypsum. Metalliferous mining was unimportant, and the only known workings, long disused, were for iron in mineralised breccias near Luccombe.

No major aquifers are present in the district, and the majority of the groundwater developments have used springs in the Hangman Sandstone. Tidal flooding is a potential problem in a few coastal areas. The hydrocarbon potential of the part of the Bristol Channel Basin that lies within the district has generally been considered to be low. However, it is now thought that Silesian strata and possible reservoir rocks (such as Triassic sandstones) may be present there (Chapter 3). Formations in the district fall into the low to medium radon potential classes (Chapter 2).

## GEOLOGY AND TOPOGRAPHY

The topography of the onshore part of the district (Figure 2, p.xii) shows a marked geological control: the Devonian rocks form hills, rising over considerable areas to over 300 m above OD and including the main areas of moorland. The younger rocks, which occupy basins that are partly fault-controlled along their northern margins, form lowland areas in the Vale of Porlock and around Minehead. The Devonian sandstones give rise to the fine rocky coastline of hogs-back cliffs, much of it difficult of access, and extensively landslipped. Between Blue Anchor and Watchet, the cliffs and foreshore platform are formed of Late Triassic and Early Jurassic mudstones and limestones, locally subject to landslipping. Coastal plains near

Porlock, and between Minehead and Blue Anchor, are occupied by Quaternary deposits — mainly saltmarsh deposits and river terrace deposits.

## OUTLINE OF GEOLOGY AND GEOLOGICAL HISTORY

Precambrian crystalline basement rocks are inferred to underlie the Bristol Channel at depths of over 6 km (Bayerley and Brooks, 1980; Mechie and Brooks, 1984). The presence of probable Lower Cambrian quartzites beneath the offshore part of the district is indicated by a poorly developed reflector on seismic profiles; a shallow shelf sea probably extended over the district during the Early Cambrian. Little is known of the history of the district during the rest of the Cambrian and in the Ordovician and Silurian, and rocks of these ages are not known definitely to occur. During the Mid- and Late Cambrian it is possible that South Wales and south-west England were the site of 'Pretannia', a landmass of presumed Precambrian igneous and metamorphic rocks (Brasier et al., 1992). Bevins et al. (1992) considered it possible that the district was the site of mudstone deposition in a shallow sea on the northern margin of 'Pretannia' during the early Tremadoc (Ordovician). From Arenig to mid-Caradoc times, it is probable that the district was again land, but during the late Caradoc transgression, black mud may have been deposited in a shallow sea on the northern flanks of 'Pretannia'. In mid-Ashgill times, black shale deposition may have been succeeded by grey shales, which possibly extended into the district. In late Ashgill times, a glacio-eustatic drop in sea level probably led to the re-emergence of land over the district (Bevins et al., 1992). During Silurian times, the district was probably largely land, except possibly in the early Ludlow when a shallow shelf sea may have transgressed into it (Bassett et al., 1992).

The oldest rocks exposed in the district comprise grey slates and sandstones of the Lynton Formation, which crop out in the core of the Lynton Anticline and are of early to mid-Devonian (late Emsian to early Eifelian) age. They were deposited in a shallow sea. The marine Lynton Formation is succeeded by the largely continental Hangman Sandstone Formation, a thick sequence of sandstones with subordinate mudstones, mainly Eifelian in age. The sedimentology of the Hangman Sandstone indicates deposition on a distal alluvial fan with ephemeral continental mudflats. The sediment was supplied from the more proximal parts of the alluvial fan during periods of high rainfall. Tunbridge (1984) suggested that the sequence represents an ephemeral stream/clay playa complex. However, the abundance of both sheet and channel sandstones suggests proximity to a sediment source, and indicates that this formation probably formed in a distal alluvial fan/continental mudflat environment. Palaeocurrents and petrology indicate that the source of the sandstones was the higher part of the Lower Old Red Sandstone in South Wales; however, coarse-grained sandstones and conglomerates within the formation may have been derived from an intermittently exposed landmass of

Precambrian and Lower Palaeozoic rocks in the Bristol Channel area.

Except for a very small offshore outlier of possible Silesian rocks about 6 km north-west of Porlock, Carboniferous rocks do not outcrop in the district, but Dinantian (Carboniferous Limestone) and overlying Silesian rocks may be present beneath much of the offshore area. This is based on the re-interpretation of a prominent seismic reflector and refractor as the top of the Carboniferous Limestone. Brooks et al. (1993), however, considered that these seismic events relate to pre-Devonian schistose and/or gneissose basement rocks. In the southern part of the district, Devonian rocks are thought to have been carried northwards on the Exmoor–Cannington Park Thrust (Chapter 3).

During the Variscan Orogeny, north–south compression resulted in folding and faulting of Devonian rocks. The folds trend mainly east-south-east and are mostly overturned to the north. A major structure, the Lynton Anticline, extends eastwards from between Glenthorne and Malmsmead to near Minehead; the Lynton Formation is exposed in its core near Malmsmead and Oare (Figure 1). A major Variscan thrust recognised in offshore seismic profiles was rejuvenated to form the Central Bristol Channel Fault Zone (Brooks et al., 1988).

A post-orogenic extensional phase followed the formation of the Variscan foldbelt. The New Red Sandstone Supergroup, which rests unconformably on the folded Devonian rocks, comprises material eroded from the Variscan mountains and laid down initially on an irregular land surface and in fault-controlled basins. The earliest deposits of this supergroup were probably laid down in the onshore Porlock Basin and consist of calcareous breccias and sandstones (the Luccombe Breccia) which formed on the distal part of an alluvial fan flanked by aeolian dunes; a thin local bed containing rounded boulders represents the deposits of a fluvial channel. The age of the basal Luccombe Breccia is uncertain: it is probably Triassic, but may be Permian or even latest Carboniferous (Stephanian). Outside the Porlock Basin, the Luccombe Breccia is absent and possible coeval deposits are the fluvial Budleigh Salterton Pebble Beds and Otter Sandstone. The presence of pebbles of Carboniferous Limestone in the Budleigh Salterton Pebble Beds indicates derivation from the north.

With the onset of the predominantly mud-dominated sedimentation represented by the Late Triassic (possibly Carnian to Norian) Mercia Mudstone Group, the separation of the Porlock and Minehead basins ended and more uniform conditions were established over the whole district. In the Dunster area, Hangman Sandstone probably formed hills in Mercia Mudstone Group times. Marginal deposits, such as the Dolomitic Conglomerate, which are well developed in South Wales and around the Mendips, are not present in the district, although some breccio-conglomerates occur near Dunster. The upper 80 m or so of the Mercia Mudstone are exposed and have been studied in detail near Watchet; they were apparently deposited in a semi-arid to arid, low-relief continental basin, with the mudstones of mixed origin as floodplain and playa deposits. Fresh to brackish-water lakes at times occupied shallow depressions on the alluvial/playa plain.

Significant amounts of silt and clay were transported sub-aerially into the basin. Many of the textures in the Mercia Mudstone were probably formed by post-depositional, pedogenic processes.

The succession from the red Mercia Mudstone Group (Triassic) to the grey Lias Group (Jurassic) records the change from a mainly continental environment of playas and supratidal sabkhas, to an epicontinental marine environment. The Rydon Member, which forms the bulk of the Blue Anchor Formation at the top of the Mercia Mudstone, contains sulphate nodules and carbonaceous mudstones, of possible algal mat origin, suggesting deposition in a sabkha environment. The onset of a marine transgression from the Tethyan province in the south is first evident in the organic geochemistry and biota of the uppermost beds of the Blue Anchor Formation, at the top of the Mercia Mudstone Group. With deposition of the Williton Member at the top of the Blue Anchor Formation, the environment became fully marine. The Westbury Formation of the Penarth Group represents deposition in a transgressive, littoral environment, with lower-energy stagnant or weakly oxygenated water. Contorted beds in the Cotham Member may have been caused by earthquake activity, and minor uplift produced conditions suitable for surface desiccation.

A thick, predominantly argillaceous, Lower to Upper Jurassic succession was deposited in the Bristol Channel area. There is no evidence that the Central Bristol Channel Fault Zone was active during sedimentation and the Bristol Channel Syncline thus appears to be a post-Jurassic feature. The Lower and Middle Lias consist of mudstones with regular limestone interbeds (Blue Lias Formation) at the base and were deposited in a shallow epicontinental sea. The succeeding Upper Lias consists mainly of silty shale and mudstone, the correlatives of the sandstones which are typical of the onshore Upper Lias from the Dorset coast to Gloucestershire. The faunas indicate a continuation of marine conditions. Above the Upper Lias, the equivalents of the Inferior Oolite and Great Oolite groups are represented by uniform, grey, marine shales. The Oxford Clay cannot be separated from the shales beneath, and appears to be lithologically similar to that seen onshore in southern England, which was deposited in a shallow sea. The youngest Jurassic strata in the district are possible Corallian sands, which may represent estuarine or deltaic conditions.

There is a marked break in the sedimentary record in the district above the Late Jurassic rocks. Any Cretaceous, Paleogene or Neogene deposits which formed there have been removed by erosion.

During the Quaternary glaciations, the district lay at the southern limits of the ice sheets, but the widespread presence of solifluction deposits (head) shows that it experienced a periglacial climate. Stages in the Quaternary evolution of the river systems are indicated by the presence of river terrace gravel at levels above the present alluvium. Present-day deposition along the river valleys is producing alluvium — mainly silt, clay and gravel. The melting of the last ice sheets resulted in a rise in sea level, the 'Flandrian Transgression', which produced marine deposits.

## PREVIOUS RESEARCH

The earliest detailed description of the geology of west Somerset was given by Horner (1816), together with a hand-coloured map which includes most of the district. On the map, the Devonian rocks ('Grauwacke'), Permo-Triassic ('Conglomerates, Sandstones, Red Rock &c.') and Jurassic rocks ('Lyas Strata') were distinguished, and some Quaternary deposits ('Marsh and Alluvial Land') were also shown. The approximate extent of the Lias outlier near Selworthy was indicated. The folding in the Devonian rocks was noted at Greenaleigh (Greenaley) and Hurlstone (Hurtstone) Point. The breccias (Luccombe Breccia of this memoir) in Porlock Vale were noted (but the term conglomerate was applied to them), and reference made to the very thick calcite veins in the breccias 'at Holt Farm' (probably Gillhams Quarry [SS 9197 4445]). Horner recognised that the conglomerates at Alcombe, included in the Budleigh Salterton Pebble Beds of this account, were quite distinct from the breccias of the Porlock Basin, and noted that they contained limestone pebbles. The limestones were believed to have been derived from the Devonian.

The broad outlines of the geology were shown on the Old Series one-inch map (1834, revised in 1839), on which a division was made into Devonian and 'New Red Sandstone'. Within the 'New Red Sandstone' of the Porlock Basin, a narrow band of 'Dolomitic Conglomerate' was mapped. The Lias of Selworthy was shown, and Lias was also indicated in the area east of Blue Anchor. Of the drift, the coastal flats between Porlock and Bossington and east of Minehead, were shown as 'Alluvium'.

De la Beche's account of the area in his celebrated report of 1839 was based largely on Horner (1816). In addition, he noted the occurrence of haematitic iron ore near Luccombe.

An account of the structure, succession and palaeontology of the Devonian rocks of north Devon and west Somerset was given by Etheridge (1867), and the structure of the Palaeozoic of west Somerset was further considered by Champernowne and Ussher (1879). Etheridge (1872) gave an account of the structure of the Watchet area with particular reference to the 'Rhaetic'.

A survey of the district on the one-inch scale was begun by J H Blake in the Watchet area in 1872–73 and continued by W A E Ussher in 1874 and 1879. No map or memoir was published, but manuscript notes for a memoir were prepared by Ussher.

Little has been published specifically on the Quaternary of the district. Godwin Austen (1865) gave an account of the 'submarine forest' beds in Porlock Bay which had earlier been referred to by De la Beche (1839).

Ussher (1875, 1876, 1877, 1878) contributed a series of papers based on the results of his resurvey of the 'New Red Sandstone' of Devon and Somerset, resulting in the first comprehensive stratigraphy of the sequence, applicable over the whole of the south-west peninsula. Of particular significance for the present district was his 1889 paper on the Triassic rocks and adjacent Devonian rocks of west Somerset, which included a coloured geological map of the area between Porlock Hill and the Quantocks. The 'New Red Sandstone' of the district was assigned to the

'Keuper or Upper Trias'. The 'Basement Beds' of the 'Keuper' included the Luccombe Breccia and the Budleigh Salterton Pebble Beds at Alcombe. The sandstones around and west of Minehead, those overlying the Budleigh Salterton Pebble Beds at Alcombe, and those resting on the Luccombe Breccia in the Vale of Porlock, were included in the 'Keuper Sandstones'. The 'Keuper Marl' represents the Mercia Mudstone of the modern classification. The Devonian Hangman and Foreland grits were still regarded as distinct formations, separated by the Lynton Beds.

Woodward (1893) provided the earliest detailed account of the stratigraphy of the Lower Lias. He divided the sequence into five units, although he considerably underestimated the thickness of the succession.

Richardson's classic paper (1911) on the 'Rhaetic' included a detailed account of the stratigraphy of the Blue Anchor to Watchet area.

Thomas (1940) carried out a detailed study, including mapping on the 1:10 560 scale, of the stratigraphy and structure of the 'New Red Sandstone' between Williton and Porlock, and demonstrated differences between the stratigraphy in the Porlock Basin and that around and east of Minehead (the 'Cleeve Lowland'), resulting from the physical separation of the two areas. His stratigraphy has generally been confirmed by the recent survey.

The Mercia Mudstone and its topmost division, the Blue Anchor Formation, have been studied by several authors in recent years. Mayall (1981) divided the Blue Anchor Formation into two members; the type area for the lower (Rydon) member was defined as the coast sections between Blue Anchor and St Audrie's Bay. Warrington and Whittaker (1984) described the Blue Anchor Formation, and defined a type section at Blue Anchor. Talbot et al. (1994) discussed the depositional environment of the Mercia Mudstone, including data from exposures at Watchet.

Tunbridge contributed a series of papers (e.g. 1978, 1981, 1984, 1986) on the sedimentology of the Hangman Sandstone of north Devon and west Somerset, which include references to localities within the district.

Whittaker (1972) described the Watchet Fault, an important post-Liassic transcurrent reverse fault crossing the coastline about 1 km west of Watchet.

Considerable attention has focused on the deep structure of the Exmoor area and the Bristol Channel, following the initial suggestion (Falcon, in discussion of Cook and Thirlaway, 1952) that the gravity gradient across the Quantock Hills of west Somerset might be associated with an underlying thrust. Bott et al. (1958) subsequently discovered the main gradient zone over Exmoor and suggested that it was due to a wedge of Devonian rocks thrust northwards over lower density Devonian or Carboniferous rocks. Studies in the offshore area using gravity, seismic refraction and seismic reflection data (e.g. Brooks and Thompson, 1973; Brooks and James, 1975; Brooks et al., 1977; Brooks and Al-Saadi, 1977; Mechie and Brooks, 1984) refined knowledge of the deep structure. New offshore seismic reflection data in the Bristol Channel provided direct evidence for a major Variscan thrust in the Palaeozoic basement beneath the Bristol Channel Syncline and thus confirmed earlier speculative arguments for major thrusting beneath Exmoor and the southern Bristol Channel (Brooks et al. 1988). The Central Bristol Channel Fault Zone was attributed to Mesozoic reactivation of the thrust.

Analysis of fault systems associated with the Bristol Channel Basin has been carried out by Dart et al. (1995), who gave details of structures in the Late Triassic and Early Jurassic rocks along the foreshore between Blue Anchor and Watchet, and by Nemčok et al. (1995).

The geology of the offshore part of the district was described by Evans and Thompson (1979) as part of a wider study of the Bristol Channel and South Western Approaches. A thickness of up to 2170 m of Jurassic strata was demonstrated in the central Bristol Channel.

The district is included in the 1:250 000 Series Bristol Channel Sheet 51°N 04°W, based on surveys carried out by BGS in the 1970s; solid geology, sea bed sediment, aeromagnetic and Bouguer anomaly maps are available. The geology of the inner Bristol Channel and Severn estuary was described by Evans (1982) in connection with the proposed Severn Barrage.

Aspects of sediment dynamics in the offshore area have been dealt with by Collins (1989) and McLaren et al. (1993).

# TWO

# Applied geology

The geological factors relevant to land-use planning and development within the district are reviewed in this chapter. The key issues are listed briefly before being considered in more detail.

## APPLIED GEOLOGICAL ISSUES

The district is mainly rural, except for the built-up areas around Minehead, Watchet and Porlock. Although there has been a long history of small-scale quarrying for local supplies of building stone, roadstone, brick clay, lime, etc., in most of the formations, there has been no major industrial development in the district, and hence no legacy of extensive contaminated and undermined ground. However, there remain many geologically related issues which need to be taken into account in land-use planning and development and these are listed below.

- *Foundation conditions*: geotechnical properties of rocks and engineering soils, weathering of mudstones, high and differential settlement, shrink-swell clays, chemical attack on concrete, compressible organic deposits, desiccated crust to marine deposits giving misleadingly high strengths

- *Slope stability*: coastal landslips, stability of artificial slopes

- *Artificial deposits (made ground, worked ground and made ground, landfill)*: unknown extent, inhomogeneous nature, backfilled quarries

- *Water resources*: groundwater supply (boreholes, springs)

- *Flood risk*: potential areas of flooding

- *Mineral and energy resources*: potential resources, hydrocarbons, past exploitation

- *Radon*: potential hazards

- *Land use, soils and agriculture*: distribution of high-grade land

- *Conservation*: Sites of Special Scientific Interest (SSSIs)

## PLANNING BACKGROUND

The broad framework for planning in Great Britain is set by national planning policies, other strategic guidelines, and Structure Plan policies. The Structure Plan sets out the strategic framework for land-use planning within a county, and the policies used to control and encourage it. Detailed policies and specific proposals for guiding and controlling the development and use of land are contained in Local Plans. The Department of the Environment issues Planning Policy Guidance Notes and Minerals Policy Guidance Notes containing advice on the preparation of structure and local plans and their implementation through development control.

The planning framework for the district is provided by the approved Somerset Structure Plan, Alteration No. 2 (1993; for the period 1993–2001); a Consultation Draft covering the period to 2011 was published in 1995. Local plan policies are set out in the Exmoor National Park Local Plan (1994; for the period 1994–2001) and the Minehead Area Local Plan (for the period 1989–1996). A consultation draft (Consultative Report) of the West Somerset District Local Plan was completed in 1995, and, when adopted, will replace the Minehead Area Local Plan. It will cover the period to 2006. The Exmoor National Park Local Plan also contains the Minerals Local Plan and Waste Local Plan for the park. The plans and policies are subject to amendment, and the reader is advised to consult the most recent versions at the appropriate offices (addresses are given in the Information Sources section of this memoir, p.112).

Planning considerations in the district are dominated by the fact that much of it lies within a National Park. The Exmoor National Park Local Plan emphasises that the preservation and enhancement of the natural beauty of Exmoor is the primary and overriding objective of the National Park Authority, and all development proposals within the park are judged against this fundamental purpose. The west Somerset coastal area, including the environs of Minehead and Watchet, is defined in the Somerset Structure Plan as a Special Landscape Area. Such areas have high landscape quality and, although the level of protection is less than that given to Exmoor, policies are designed to ensure that the character of these areas is safeguarded.

The West Somerset Local Plan (Consultative Report, 1995) indicates that the principal direction of expansion of Minehead in the last 15 years has been eastwards, because of the environmental restraints restricting growth of the town in other directions and because of the need to ease traffic congestion in the town itself. Development has included the construction of Seaward Way, the Minehead Enterprise Park, and housing. This development impinges on the saltmarsh areas east of Seaward Way and Butlin's Holiday Camp and, in order to provide some planning protection, this area was included within the Special Landscape Area by amendment of the approved Somerset Structure Plan. The West Somerset District Local Plan Consultative Report (1995) identifies land for residential development during the life of the plan in areas adjacent to and west of Seaward Way. Any housing development would have to take into account the geotechnical and drainage problems associated with the saltmarsh deposits, outlined in the foundation conditions section, below.

Planning factors in the district are shown on Figure 3.

## FOUNDATION CONDITIONS

The suitability of bedrock and superficial deposits for the construction of foundations is dependent mainly on their geotechnical properties. The stability of engineering structures is also related to geological factors, including local geological structures and slope stability. The rock and engineering soil types of the district cover a range of geotechnical materials, mainly in the sandstone to mudstone range, and their weathered products. There are few available geotechnical data for the geological formations of the district. The comments which follow are intended as a general guide to the likely behaviour of the materials encountered in the area, but may not take into account local variations in lithology or behaviour of the formations described.

The geotechnical account below is not intended as a substitute for an appropriate site investigation (Anon., 1981) which should be carried out before the commencement of any engineering works. The BGS holds 1:10 000

scale geological maps and other geological data as well as borehole and geotechnical data which are of value for assessing the suitability of a site for development.

### Lynton Formation

The Lynton Formation comprises grey slates with subordinate slaty sandstone and slaty siltstone. No geotechnical information is available. The lithologies are likely to be similar in engineering behaviour to the equivalent lithologies in the Hangman Sandstone and, in an unweathered condition, to offer good foundation conditions with good bearing capacity and low settlement. Their excavation may require ripping or the use of hydraulic or pneumatic rock breakers, depending on the degree of fracturing and the state of weathering of the material. The excavated material may be suitable for fill if it is reduced to a suitable size grading, and if mudstone is not present in significant amounts.

### Hangman Sandstone Formation

The Hangman Sandstone comprises purple, grey and green, generally medium to thickly bedded, strong to very strong sandstone with very thin to thin, cleaved mudstone and very fine-grained sandstone interbeds. Units of

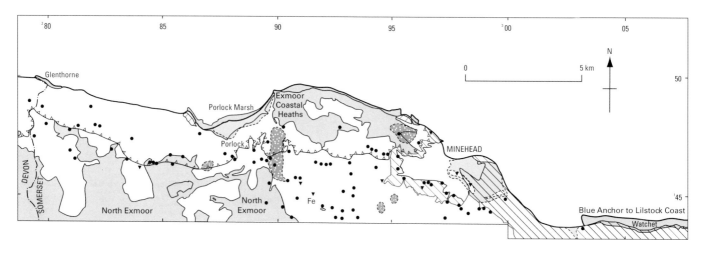

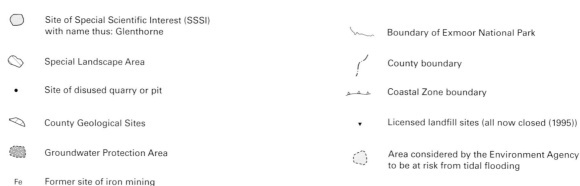

**Figure 3**  Planning map of the district.

purple, thickly bedded, cleaved mudstone with thin, very fine-grained sandstone interbeds are found in some parts of the sequence. The beds are folded along east-northeast axes and are highly fractured, but the fractures are mostly closed and tight.

The formation is likely to offer good foundation conditions with high bearing capacity and low settlement. Its excavation is likely to require ripping or the use of hydraulic or pneumatic rock breakers, depending on the thickness of the bedding, the degree of fracturing, the tightness of fractures and the state of weathering. Blasting may be necessary to loosen the tightly closed fractures, or to break up thickly bedded intact material. The excavated material may be used as fill if reduced to a suitable size grading.

## Luccombe Breccia Formation

The Luccombe Breccia comprises reddish brown, well-cemented, commonly calcareous, sandy breccia, and sandstone. The breccia clasts are angular and platy, and comprise slate, shale and sandstone. A 'boulder bed' of cobbles and boulders, up to 500 mm in diameter, of strong to very strong sandstone is present within the breccia (Plate 20). No geotechnical data are available for the formation.

The formation is expected to offer good foundation conditions with a high bearing capacity and low settlement. Its excavation may prove difficult in the coarser breccia and 'boulder bed,' and may require the use of hydraulic or pneumatic rock breakers. The excavated material may be usable as fill.

## Budleigh Salterton Pebble Beds

The Budleigh Salterton Pebble Beds consist of reddish brown conglomerate containing rounded to subrounded cobbles and boulders of limestone and sandstone up to 300 mm diameter in a calcareous sandstone matrix, interbedded with sandstone containing some medium to coarse gravel. The formation has only a very small outcrop within the district and no geotechnical information is available; its engineering behaviour is likely to be broadly similar to that of the coarser parts of the Luccombe Breccia.

## Otter Sandstone Formation

The Otter Sandstone comprises reddish brown, yellowish brown, or green, locally mottled, fine- to medium-grained, medium to thickly bedded, locally moderately cemented, weak to moderately weak sandstone. Some thin beds of brown mudstone and breccia are present. The formation is expected to offer good foundation conditions with good bearing capacity and low settlement.

Excavation of the Otter Sandstone may be possible by digging in the poorly cemented or weathered material, but may require ripping or the use of hydraulic or pneumatic rock breakers in the thicker-bedded, better-cemented material. The excavated material may be usable as fill.

## Mercia Mudstone Group (excluding the Blue Anchor Formation)

In the eastern part of the district, including part of the town of Minehead, the red Mercia Mudstone comprises reddish brown, dolomitic, slightly calcareous mudstone with some minor beds of limestone and sandstone. In places, the mudstone contains nodules of, and is cut by, veins of gypsum. Near Dunster [SS 9940 4385] a moderately well-cemented breccio-conglomerate with boulders of Devonian sandstone occurs within the mudstones.

In the western part of the district, sandstones are interbedded in the lower part of the red Mercia Mudstone. These comprise reddish brown, well-cemented, medium- to coarse-grained, moderately weak to moderately strong sandstone.

The mudstone weathers initially to a blocky, friable material, and ultimately to soft to stiff, silty, sandy, fissured, low-plasticity clay.

In an unweathered state, the red Mercia Mudstone Group mudstones may be expected to offer good foundation conditions with good bearing capacity and low settlement. However, attention should be paid to the weathering grade of the material, since the bearing capacity and settlement will be impaired as the weathering state advances. Sandstones and breccias in the group may be expected to offer good foundation conditions in an unweathered state, and the sandstones will be less affected by weathering than the breccias in which the mudstone clasts may suffer from softening. Where mudstones, breccias and sandstones are present on a site, consideration should be given to the possibility of differential settlement over the different material types.

Excavations in weathered mudstones may be expected to be accomplished by digging, with the possibility of ripping in the unweathered material. Excavation of the better-cemented and thicker-bedded sandstones and breccias may require ripping or the use of hydraulic or pneumatic rock-breaking machinery. The coarser material may be suitable for use as fill, but use of the more cohesive material as fill may be subject to the moisture content of the material being suitably controlled.

## Blue Anchor Formation

The Blue Anchor Formation is the highest formation in the Mercia Mudstone Group and comprises grey and green mudstones with gypsum veins and beds. No geotechnical information is available for the formation in the district, but its engineering behaviour is probably comparable to that of the red mudstone facies of the Mercia Mudstone.

## Penarth Group

The Penarth Group comprises the Westbury and Lilstock formations and includes a wide range of generally thinly laminated to thinly bedded materials with contrasting geotechnical properties. These include pale grey, grey, very dark grey, black, and greenish grey, shaly, calcareous, pyritic mudstones, and sandy limestones.

No geotechnical information is available from the district for these materials, but it is likely that some, if not most, may be weak, contain high-plasticity clays and may weather easily, giving rise to both swelling clays and, with the breakdown of iron pyrites, the generation of acidic, sulphate-rich solutions. Thus, the Penarth Group may offer poor ground conditions with low bearing capacity, high settlement and possibilities of clay shrinkage or swelling, chemical attack on buried concrete structures or service connections, and slope instability on natural or cut slopes. The expected wide range in geotechnical properties may give rise to differential settlement where a foundation crosses from one material to another.

## Blue Lias Formation

No geotechnical data are available from the Blue Lias of the district. The formation comprises grey, calcareous, very thinly to thinly bedded, weak mudstone with thinly bedded, shelly, moderately strong to strong limestone. The mudstone weathers to high-plasticity clay. Unweathered Blue Lias may offer reasonable bearing capacity, but weathered material may be softened and offer low bearing capacity with high settlement. In each case, foundations may need to be designed to take account of the shrinking and swelling of clays associated with changes in water content.

## Marine deposits (onshore)

In the near-surface zone the marine deposits may form a crust, approximately 1 to 2 m thick, of oxidised, desiccated, pseudo-overconsolidated, brown to grey, very stiff to stiff, becoming firm with depth, closely fissured, silty clay of high plasticity. The material below the crust may be soft, normally consolidated, blue-grey, silty clay. Organic or peaty deposits may be present within the silty clay. Dense, brownish red gravel or sand may be present below the soft, silty clay.

The marine deposits are of generally low bearing capacity, and construction on them should take into account the lower strength and higher consolidation characteristics of the material below the stronger, desiccated, crust. Settlement may be uneven owing to the presence of local deposits of highly compressible peaty or organic material. Care should be taken to avoid site traffic cutting through the crust to the detriment of vehicle mobility. The use of temporary roads constructed of suitable fill placed on a geotextile membrane may be advisable.

Excavations in the marine deposits should be possible by digging, but immediate support of the excavation sides may be necessary, particularly in the softer material. The use of selected excavated material as fill may be possible, but the soft and organic material is unlikely to be suitable.

## River terrace deposits

No geotechnical data are available for the river terrace deposits in the district. They comprise mainly gravel and represent the dissected remains of former alluvial deposits at levels above the present river floodplain. The river terrace deposits are rarely exposed, but temporary exposures show them to be brown, poorly bedded, fine-medium- and coarse-grained gravel with subrounded to subangular sandstone clasts, and some cobbles and small boulders, in a matrix of brown, silty sand.

The thickness of the river terrace deposits varies from place to place. Thicknesses of up to 4.6 m were proved in excavations between Loxhole Bridge [SS 9960 4395] and Sea Lane End [ST 0052 4445] and geophysical soundings have indicated thicknesses of 22 m and 34 m north of Porlock [SS 8857 4788 and SS 8869 4736 respectively].

The deposits are likely to offer good foundation conditions with low settlement. Excavation can generally be achieved by digging, but immediate support may be necessary in loose material and de-watering may be necessary if the water table is reached during excavation. The excavated material is likely to be suitable for use as fill.

## Alluvium

No geotechnical data are available for the alluvium, which forms narrow deposits along the bottoms of the small valleys in the west of the district, the broader area along Horner Water near West Luccombe [SS 899 463] and the broad spread between Tivington Knowle [SS 926 446] and Holnicote [SS 908 460]. However, it is expected from knowledge of alluvium in other places, that it will comprise normally consolidated gravel, sand, silt and clay, and combinations of them, of low strength. Peat and organic-rich material may be present as planar beds and/or lenticular bodies within units of other material.

Alluvium is expected to offer poor foundation conditions of low bearing capacity and high settlement. Settlement may be uneven where a foundation is placed across material of contrasting geotechnical properties. A desiccated crust of higher bearing capacity may be encountered overlying weaker material. Excavations are expected to be possible by digging, but immediate support may be needed to maintain the excavation in a stable condition. Excavations may require de-watering as the water tables are likely to be close to the ground surface. The coarser excavated material may be usable as fill, but the soft and organic material is unlikely to be suitable.

## Head

Head deposits in the district vary in thickness, but may be several metres thick at the bottom of slopes. The geotechnical properties of head are dependent on those of the parent material but, in general, the deposits are in a remoulded or loose condition, and are weak and compressible.

## SLOPE STABILITY

Natural slopes inland appear to be generally stable, but major landslips, some of which are currently active, are present on the coastal slopes. The slips involve the Hangman Sandstone west of Minehead, and the Mercia Mudstone, Penarth Group and Blue Lias between Blue Anchor and Watchet. Artificial slopes may be prone to instability depending on the formation excavated, local groundwater conditions and other factors.

Coastal landslips are widely developed on the Hangman Sandstone, and some are currently active (Plate 1). Inland slips are rare. The main areas of former and present instability are on the coastal slope from west of Porlock Weir to Glenthorne, and between Hurlstone Point and Greenaleigh (Figure 36). West of Porlock Weir, recent slips in Yearnor Wood have affected the South West Coast Path, and consequently investigations into the cause of the instability have been carried out (Anon., 1992, 1994).

The northward-facing slopes formed by the Hangman Sandstone along the coastal section between the western margin of the map [SS 787 500] and Worthy [SS 858 482] show evidence of extensive landsliding in Hangman Sandstone head and Hangman Sandstone bedrock (Anon., 1992, 1994). The slopes dip seaward at about 35 to 45°, but may be locally steeper (50 to 60°) on the lower slopes. The slopes are generally thickly wooded, and include Yearnor, Culbone and Embelle woods. In August 1994, there was much evidence of past and current landslide activity on the slopes in the form of degraded back scarps, rockfalls, scree and hummocky ground, but the thickness of the vegetation made it difficult to establish a clear pattern of past activity.

At the back of the beach west of Porlock Weir, wave action has eroded the head deposits from the base of the slope which has, in turn, removed support from the deposits farther upslope. In August 1994 there was an active, shallow, translational landslide in Yearnor Wood (Plate 1) where the slope had failed progressively upslope, in a series of rock block slides, debris slides and flows. The slope-failure had progressed approximately 200 m upslope, cutting the lower coastal path. The highly fractured sandstone bedrock had failed by a series of wedge failures, which were controlled by seaward-dipping bedding planes and high-angle jointing, whose intersection dipped downslope at an angle approximately coincident with that of the slope. Large block slides on bedding planes whose dip was approximately coincident with the hill slope were also present higher up the slope. The initial failure appears to have taken place on thin mudstone beds between the thicker sandstone beds.

The sandstone is highly fractured and the failed masses have broken up into debris flows as they moved down the slope.

The topography of the coastal section between Hurlstone Point [SS 900 492] and Greenaleigh Point [SS 956 482] is also affected by past landsliding processes, and much of the slope shows a degraded, hummocky topography of complex landsliding involving debris slides, flows, rockfalls and rotational failures. Currently, these slopes are largely covered by rough grass, scrub and rock debris, and no active landsliding was observed at the time of the survey. In the coastal landslipped area south-east of Minehead Bluff, called the Western Brockholes, 8 to 10 m-high crags [SS 9163 4907] form the backscar to an extensive area of slipped masses of shattered sandstone forming blocks and ridges separated by areas of angular sandstone boulder scree.

The processes which control the instability or stability of the seaward slopes formed by the Hangman Sandstone are complex, but include the following factors:

1. The progressive weathering and weakening of the mudstone interbeds.
2. The relationship between discontinuities (bedding, jointing and faulting) and the dip of the hill slope.
3. Changes in land use and vegetation.
4. High and/or intense rainfall periods with a consequent rise in groundwater levels
5. Basal erosion due to extreme high tides and storm surges.

The stability problems are severe, owing to the seaward-dipping component of the bedding and the highly fractured nature of the rock. There is difficulty in determining whether the slips are in head which is moving on the underlying bedding surfaces as debris slides, or whether movement starts as a wedge failure/block slide, moving

**Plate 1**  Coastal landslips [SS 853 484] in Hangman Sandstone bedrock and head, Yearnor Wood, west of Porlock Weir (July 1995) (GS478).

on a shale bed in the sandstone, which immediately breaks up into a debris slide/flow. Probably both mechanisms are involved.

The primary cause of recent landslipping west of Porlock Weir has been attributed to the removal of previously slipped debris from the foot of the steep coastal hillslope, assisted by wave action, currents and tides (Anon., 1992, 1994). Phases of enhanced instability are possibly related to particularly high tidal levels and very infrequent storm-tidal surges; they occur perhaps every 10 to 100 years and are capable of initiating large-scale instability. Extreme high water levels were inferred to have occurred at Porlock in 1607, 1703, 1873, 1910 and, more recently, in December 1981 and February 1990. It is significant that only one high-level event occurred in the century between 1880 and 1980, while two such events have already occurred within the last decade. The earlier of these two tidal surges correlates quite well with the onset of recent instability in the early 1980s (Anon., 1992, 1994).

Although no confirmed evidence of landsliding was observed on the inland outcrop of the Hangman Sandstone, it is possible that, where it has developed a thick mantle of head and the bedding coincides with the dip of the hill slope, engineering work which causes removal of material from the base of a slope may trigger landsliding in a similar manner to that observed on the coast. Therefore, consideration should be given to slope instability in all operations involving the removal or deposition of material on slopes of Hangman Sandstone.

The stability of cut slopes in the Luccombe Breccia may be expected to be good, subject to the absence of structural discontinuities dipping out of the face, and the possibility of boulders or cobbles dropping out of the face.

The stability of cut faces in the Otter Sandstone may be good, as shown by the near-vertical exposure of thick, cross-bedded sandstone and fine-grained breccia in Holloway Street, Minehead [SS 967 464]. When cut to a lower angle, as for example in a cutting [SS 948 457] on the A39 road near Bratton, the Otter Sandstone may weather to sandy rubbly head and become vegetated, with only the better-cemented beds remaining exposed.

The instability affecting the Mercia Mudstone, Penarth Group and Blue Lias along the coast from Blue Anchor to Watchet is particularly complex due to faulting which has brought several landslide-susceptible lithologies into close association and has, itself, provided planes along which sliding may occur. The

Mercia Mudstone stands at steep angles in the cliffs, but is prone to undercutting by wave action and may fail by rock fall or block slide along discontinuities which dip out of the cliff face. In some places, the head which locally caps the cliff section may fail by mud or debris flows down the cliff face, depending on the water content of the material.

The Blue Anchor Formation is well exposed at its type section at Blue Anchor and forms the cliff eastwards towards Watchet. Although it contains much nodular, bedded and vein gypsum, it is sufficiently strong to sustain a steep cliff profile. However, there is much faulting of the ground along this section of the coast, and some large-scale block slides have occurred along major discontinuities which dip seaward at medium to high angles. The bedding of the formation dips seaward in some parts of the cliff and offers the possibility for sliding failure to take place on bedding planes. The formation is subject to undercutting by wave action at the base of the cliff, and solution of the abundant gypsum is also pronounced, both processes potentially leading to rockfall. Substantial rockfalls occurred in the winter of 1995/1996, affecting mainly the Blue Anchor Formation in the vicinity of Blue Anchor Cliff [ST 0385 4368] (Plate 2).

Where the Penarth Group is exposed in the cliffs west of Watchet, it is associated with complex landsliding. Inland, natural or cut slopes may give rise to instability.

In many areas, Lias Group clays and superficial materials derived from them are associated with slope instability, and the possibility of landslipping should be considered prior to engineering activities on such slopes in the district. Near Watchet, coastal landslips are related to steeply seaward-dipping Lias and to local groundwater conditions (Grainger and Kalaugher, 1996). Where the Blue Lias forms the cliff at Warren Bay, west of Watchet, it is associated with complex landsliding involving rotational failures, mudflows and mudslides.

**Plate 2** Blue Anchor Cliff [ST 0385 4368] showing substantial rockfalls affecting mainly the Blue Anchor Formation (March 1996) (GS479).

In periglacial conditions, slopes covered by cohesive, plastic head may develop shear surfaces during solifluction. Although the surface expression of past movement may have been degraded, relict or fossil shear surfaces may still be present. It is possible for such shear planes to be reactivated and slope movements to resume if an increase in porewater pressure is caused by rising groundwater levels, or if the slope is undercut by the removal of material from the base of the slope.

## MADE GROUND, WORKED GROUND AND MADE GROUND, LANDFILL

*Made ground* consists of artificial material deposited on the original ground surface, and *worked ground and made ground* comprises artificial deposits filling disused pits and quarries. The identification of these deposits is an important aspect of land-use planning. Apart from presenting potential problems in terms of foundation conditions, there are implications for pollution of groundwater and in the fields of agriculture and forestry. Major problems with deposits of domestic and industrial waste stem from their low strength and inhomogeneous nature. Large structures built on such deposits may require special foundation designs to overcome possible excessive and differential settlement. Such problems may be particularly important at the boundary of infilled land where structures straddle an area with contrasting ground conditions. Land that is returned to agriculture or forestry after infilling may not be in the same condition as the surrounding natural ground. Differences in drainage characteristics and the nature of the restored upper soil layer may variously affect the type and yield of crops. Attempts to afforest areas of landfill may be affected by the penetration of deep root systems into layers of buried domestic or industrial waste.

The disposal of toxic substances in *landfill* sites can present special problems affecting the environment. Such waste can be considered in two categories, firstly, material which is inherently toxic and, secondly, material which may react to generate toxic or hazardous substances. Examples of the former are a wide range of industrial chemical wastes such as mineral acids, cyanide residues and phenols. Some materials rich in cellulose, such as domestic refuse, can break down by the action of bacteria to form methane, or undergo combustion to form gases rich in carbon monoxide, and these examples would fall into the second category

The potential pollution of aquifers by leachate from waste disposal sites should be considered as a factor in the licensing of sites.

Site investigation programmes should be designed to take into account the problems outlined above and to probe the thickness, lateral extent and composition of artificial deposits. This is particularly important where waste disposal has occurred prior to the licensing regulations of 1976.

Licensed waste disposal (landfill) sites are a category of made ground or worked ground and made ground. There are eight such sites within the district, all of which are now closed (Table 2). Four were located on the Hangman Sandstone and two on the Mercia Mudstone. These contain domestic and inert waste, and leachate

**Table 2** Waste disposal sites in the district.

| Name | Stratigraphy | National Grid reference | Type of waste | Dates in use | |
|---|---|---|---|---|---|
| | | | | From | To |
| Middlecombe, Minehead | Hangman Sandstone | SS 946 455 | Construction and demolition associated with highway works | 1974 | 1978 |
| Mart Road, Minehead | Marine deposits | SS 978 458 | Natural excavated material | 1988 | 1991 |
| Higher Hopcott Quarry | Hangman Sandstone | SS 963 452 | Wood, soil, metal, hardcore | 1987 | 1993 |
| Rydons Quarry, Luccombe | Sandstone in Mercia Mudstone | SS 9100 4555 | Builders' rubble, subsoil, topsoil, timber | 1989 | 1992 |
| Minehead Relief Road | Marine deposits | SS 984 453 | Weathered tarmac, concrete, natural excavated material, subsoil, topsoil | 1990 | 1992 |
| Pittcombe Head, Porlock | Hangman Sandstone | SS 841 462 | Domestic, builders' waste, soil, rubble | 1986 | 1990 |
| Pinspit Quarry, Luccombe | Mercia Mudstone | SS 916 449 | Domestic | 1965 | 1972 |
| Culver Cliff, Minehead | Hangman Sandstone | SS 967 475 | Domestic | 1940s | 1972 |

transport is limited by the low permeability of the rocks. Two sites are located on the marine deposits south-east of Minehead. Only inert waste has been licensed for disposal owing to the sensitive nature of these sites which are either close to or below the water table, and are associated with highly permeable rocks.

The Waste Local Plan for the Exmoor National Park forms part of the Exmoor National Park Local Plan. The policy is that new waste disposal facilities should not be located within the National Park.

The most extensive area of made ground shown on the map is that centred on Butlin's 'Somerwest World' Holiday Camp [SS 985 460], which is reported to have been formed of quarry hardcore dumped over the saltmarsh deposits in 1961–62. The source and thickness of the material is unknown. In the urban areas of Minehead, Watchet and Porlock, made ground has not been shown on the map because of the difficulty of identifying it, but is likely that there are areas of artificial deposits of unpredictable nature, extent and thickness within these urban areas.

A tip [SS 8933 4838] near Bossington, behind the shingle ridge and adjacent to the place where the Horner Water meets the ridge, consists of up to 3 m of made ground within which were noted builders' rubble and household waste. An elongate area of made ground [centred on SS 967 475] north-west of Minehead Harbour, of uncertain composition, overlies probable saltmarsh deposits, and is bounded on the seaward side by a ridge of storm beach shingle. Other areas of made ground include the material used in the construction of railway embankments for the West Somerset Railway in the Blue Anchor area, possible sources of which were the cuttings east of Blue Anchor Station. Near Watchet, an old tip, which possibly included household waste, was sited at Daw's Castle [ST 0615 4321], on the cliff top close to the line of the Watchet Fault.

Worked ground and made ground is of minor extent in the district. Most old pits and quarries contain variable, usually minor, amounts of waste as a by-product of quarry working. A few quarries and pits have been wholly or partly filled with imported material, the nature of which is generally unknown. North-east of Marsh Street, an old pit [SS 999 447] in gravelly river terrace deposits has been partly infilled. Pinspit [SS 9163 4492], a former claypit in the Mercia Mudstone near Luccombe, has been almost completely infilled, as has Rydons Quarry [SS 9100 4555].

## GROUNDWATER RESOURCES

The Minehead district is included mainly within Sheet 42 (Somerset Coast) of the Environment Agency Groundwater Vulnerability Map Series (1:100 000 scale, prepared by BGS in collaboration with the Soil Survey and Land Research Centre). On this map, the Environment Agency has classified the Budleigh Salterton Pebble Beds and Otter Sandstone as major aquifers. The Mercia Mudstone Group (including the Blue Anchor Formation) is classified as a non-aquifer; the remaining units are classified as minor aquifers. The distribution of low-permeability drifts

occurring at the surface and overlying major and minor aquifers is also indicated on the map; in this district, these low-permeability deposits are mainly saltmarsh deposits.

Groundwater Protection Areas, defined by the Environment Agency, are identified on the Proposals Map of the Exmoor National Park Local Plan (1994) in order to prevent development which would cause groundwater contamination within these zones. The areas are sited around the public water supply boreholes in Moor Wood (Minehead), between Bossington and Horner, around the springs in Periton Hill Plantation and about 1 km south of West Porlock (Figure 3).

Apart from the Budleigh Salterton Pebble Beds and Otter Sandstone, which underlie a small part of the district, there are no regionally significant aquifers. Most of the district is underlain by the Devonian Hangman Sandstone, which is well cemented and of low permeability. The formation offers a limited groundwater resource potential from fissures which give rise to springs. Some Quaternary sand and gravel deposits also form useful local aquifers.

Groundwater supplies are typically small, but they satisfy private domestic use. The BGS and the Environment Agency hold information on 46 groundwater sources, comprising 26 boreholes or wells, with depths mostly in the range 25 to 35 m, and 20 springs that are used for water supplies. In general these sources draw from the Hangman Sandstone, the Otter Sandstone and the Mercia Mudstone.

### Lynton Formation

Weakly permeable, well-cemented, fine-grained rocks of the Devonian Lynton Formation underlie a small area in the west of the district. One borehole [SS 8036 4722] near Oare yielded 0.8 litres per second ($1 s^{-1}$), for a small drawdown.

### Hangman Sandstone Formation

The Hangman Sandstone is composed of sandstones with subordinate mudstones. There is some inter-granular permeability in the sandstone sequences, and some groundwater storage is available in fissures (Richardson, 1928). In general, the fissures are not well connected and this restricts groundwater flow and limits ground-water potential.

A spring [SS 935 438] north-west of Wootton Courtenay is located at the faulted junction between the Hangman Sandstone and the Mercia Mudstone. In general, yields from springs range between 0.1 and $1.9 1 s^{-1}$ with an average of about $0.7 1 s^{-1}$. Most springs show a significant seasonal variation in flow.

Borehole yields range from 0.2 to $9.0 1 s^{-1}$. However, yields of about $0.5 1 s^{-1}$ (British Geological Survey, 1982) are more typical. Most boreholes and wells range from 13 to 60 m deep with drawdowns from 7 to 26 m. Yields are largely dependent on the number and the size of fractures penetrated. Initial yields may drop sharply with time due to the slow rate of recharge from the surrounding rock mass. A seasonal fall in yield may also be experienced in late summer.

The majority of the groundwater developments in the district have utilised springs in the Hangman Sandstone (Richardson, 1928). Some of these are associated with faults within the formation, and others are associated with junctions and faulted junctions with younger rocks. Public water supply boreholes for Minehead at Moor Wood [SS 958 473] have a licensed abstraction of 909 cubic metres per day ($m^3 d^{-1}$); there are two boreholes 180 and 300 mm in diameter and 31 and 46 m deep respectively, and they yield 4 and 9 $l s^{-1}$ for drawdowns of 7 and 26 m respectively. The water supply for Minehead also utilises springs [SS 946 446 and 952 449] in Periton Hill Plantation, which have a combined licensed abstraction of 95 $m^3 d^{-1}$. One spring flows at between 0.3 and 0.7 $l s^{-1}$. Porlock's water supply utilises a spring [SS 872 463] at the head of Allerpark Combe, which has an abstraction licence for 136 $m^3 d^{-1}$.

The groundwater quality of the public supply boreholes at Moor Wood is good, with a total dissolved solids (TDS) content of 402 milligrams per litre (mg $l^{-1}$) and a pH of 6.3. It has a total hardness of 88 mg $l^{-1}$ (as $CaCO_3$) and a chloride ion concentration of 53 mg $l^{-1}$. The sulphate ion concentration is 29 mg $l^{-1}$, sodium ion concentration is 25 mg $l^{-1}$ and the calcium ion concentration is 21 mg $l^{-1}$. One public supply spring in Periton Hill Plantation has a low TDS of 66 mg $l^{-1}$ and a pH of 6.9. It has a total hardness of 35 mg $l^{-1}$ and a chloride ion concentration of 31 mg $l^{-1}$.

### Luccombe Breccia Formation and Budleigh Salterton Pebble Beds

Little is known of the hydrogeological characteristics of either formation in this district, and there does not appear to have been any groundwater abstraction from them.

### Otter Sandstone Formation

The porosity of the Otter Sandstone at Somerset localities outside the district is locally relatively low, between 5 and 10 per cent (Sherrell, 1970). The intergranular permeability in the sandstones can be important where they are weakly cemented and/or the silt and clay content is low. However, the secondary permeability derived from joints and fissures controls the hydraulic properties of the sandstone (Sherrell, 1970). Beds of mudstone may divide the rocks into a multi-layered aquifer. Borehole yields are dependent on the presence of water-bearing fissures. These rocks usually behave as a leaky confined aquifer with a storativity of $10^{-3}$ to $10^{-4}$ (Sherrell, 1970). Faults mostly act as barriers to flow due to their well-cemented nature, although some act as recharge boundaries.

A water-investigation borehole (Holloway Street Borehole, p.64) was drilled at Minehead [SS 9673 4637] to a depth of 84 m in the Otter Sandstone in 1967, and later abandoned. The lower part of the sequence contained mudstone, clay and silt. The yield was only 6.3 $l s^{-1}$ and the transmissivity was 21 $m^2 d^{-1}$ (Sherrell, 1970). After 6.5 hours of pumping the rate of drawdown accelerated, indicating that the aquifer is of limited extent. This probably relates to the faulted boundary with the Hangman

Sandstone, 270 m to the west. This limitation may also be found elsewhere, due to the restricted size and faulted nature of the outcrops.

A well field of six shallow boreholes between 15 and 30 m deep has a combined abstraction license of 3820 $m^3 d^{-1}$ at a holiday camp [SS 985 460] east of Minehead. This site exploits the Otter Sandstone which underlies between 8 and 10 m of saltmarsh deposits, which are cased out. The elevation of the coastal plain is about 5 m above OD (Ordnance Datum), and boreholes are as close as 400 m from the coast. The boreholes have been spread out in order to avoid excessive drawdown during pumping which could cause the intrusion of saline groundwater. Yields from the individual boreholes range from 4 to 8 $l s^{-1}$. The rest water level is about 3 m above OD and drawdown is limited to about 2 m. Where the overlying marine deposits are arenaceous there is likely to be induced recharge, as a result of pumping, from these rocks. There are no other examples in the district of groundwater abstraction from the Otter Sandstone.

Characteristically the groundwater is a calcium, magnesium bicarbonate water. Groundwater from the Holloway Street Borehole had a pH of 7.2, a total hardness of 215 mg $l^{-1}$ and a chloride ion concentration of 40 mg $l^{-1}$. At the holiday camp the groundwater had a pH of 7.3 and a total hardness of 180 mg $l^{-1}$. The chloride ion concentration had a range from 45 to 165 mg $l^{-1}$, which may reflect the proximity of the coast.

### Red Mercia Mudstone Group

The red mudstones of the Mercia Mudstone are of low permeability, but they include numerous small fissures which allow the passage of water (Richardson, 1928). Thin beds of sandstone are also present, together with some veins of gypsum. The sandstone horizons can form useful aquifers under favourable circumstances. These may occur where the Mercia Mudstone overlies older water-bearing rocks, in particular the Hangman Sandstone (Richardson, 1928). The Mercia Mudstone is generally regarded as an aquiclude with some permeable layers. The sandstone beds are usually well cemented and the groundwater resource is largely confined to fissures and joints within the rock.

The success of a borehole is usually dependent on the thickness of sandstone penetrated and the degree of fracturing present. Small yields from 0.1 to 1.5 $l s^{-1}$ (British Geological Survey, 1982) are derived from boreholes in these rocks. Owing to the presence of gypsum veins, the groundwater quality is typically hard, with a total hardness of 400 to 500 mg $l^{-1}$.

### Blue Anchor Formation, Penarth Group and Blue Lias Formation

The Blue Anchor Formation, Penarth Group and the Blue Lias Formation are largely composed of mudstone, although beds of limestone are present in the Penarth Group and the Blue Lias Formation. Thus, they are largely aquicludes and not important for water supply purposes, although fissured limestone may provide small yields of hard water to some large diameter wells. Springs may be present where beds of mudstone and interlayered limestone outcrop.

## Quaternary deposits

Marine deposits, river terrace deposits and alluvium occur in the broad valley to the north and east of Porlock and on the coastal platform at Minehead and to the south-east. Alluvium also occurs as a narrow strip in the base of the main valleys in the upland region. The marine deposits include highly permeable sequences of sand and gravel. Sands and gravels also form an important component of the river terrace deposits and are commonly found at the base of the alluvium beneath silt and clay. Most of these sequences have a high porosity and a thickness limited to a few metres, of which only the lower part is saturated. The high permeability is based on the intergranular storage and flow of groundwater and is usually associated with low hydraulic gradients. In the alluvium and marine deposits these aquifers are commonly in hydraulic continuity with the local streams and rivers. This may also be the case with the terrace deposits where the elevation of the water table is coincident with the local river. Where these deposits are adjacent to the coast they can be in hydraulic continuity with the sea if the base of the permeable sequence is below sea level. Under natural conditions the hydraulic gradient will be towards the coast. The thin alluvial deposits in the uplands are of little hydrogeological significance and are dominated by the effect of the underlying rocks, mostly the Hangman Sandstone, and the hydrology of the associated streams and rivers.

In the upland areas, the sands and gravels in the alluvial deposits are rarely thick enough to provide perennial sources for water supply. The groundwater resources can be utilised in the broad lowland deposits, but they may be vulnerable to saline intrusion near the coast if pumping water levels fall below sea level. Where the groundwater source is in hydraulic connection with the local river, the quality of water is likely to be similar. Also, due to the near-surface character and high permeability of these deposits, they can be easily polluted by surface sources. There have not been any significant developments of this resource, which may reflect the difficulties involved. Indeed they are usually cased out when developing an underlying aquifer as, for example, at the Minehead holiday camp.

The seven Environment Agency licensed abstractions from rivers in the district range from 11,232 m³ d⁻¹ from the Horner Water at Lynch [SS 899 478] to 27 m³ d⁻¹ from a stream in the Minehead holiday camp, and have a median of 246 m³ d⁻¹.

## FLOOD RISK

Areas considered to be at risk from flooding are identified on maps prepared by the Environment Agency, in accordance with the requirements of the Water Resources Act (1991). The areas at risk within the district are shown on Figure 3; these include the area of saltmarsh deposits behind the shingle ridge of Porlock Bay, an area of mainly saltmarsh deposits east of Minehead and an area at the mouth of the Pill River, Blue Anchor.

Should sea level rise, such coastal areas would be at increasing risk of flooding. However, long-term climatic changes and their effects on sea level are not yet fully understood. For example, there is evidence that the Bristol Channel area is subsiding (or sea level is rising) at a rate of between 0.2 and 0.5 mm/year (Shennan, 1992) (compared with a global rise of 1.5 to 2 mm/year); however, at Milford Haven the tide gauge data over the past 12 years shows that sea level has fallen at a rate of 3.40 to 1.7 mm/year. The reasons for this anomalous fall are not fully understood (Shennan and Woodworth, 1992).

## MINERAL AND ENERGY RESOURCES

### Construction materials

Many small quarries and pits, scattered over the district, formerly exploited most of the formations for varied purposes, including the extraction of building stone, roadstone, brick clay, gravel, lime and gypsum, but there are now no active quarries in the district. Details of selected quarries are given by Edwards (1996). Many of the quarries contain small resources of local building stone. The Exmoor National Park Minerals Local Plan (1994) policy is that proposals for mineral extraction in the park would be subject to the most rigorous examination and not permitted unless there was an overriding national need. However, it was considered that small-scale reworking of disused quarries could be permitted in order to provide supplies of local building stone, provided that such use had no adverse impacts, and that the site was restored after the cessation of working.

SAND AND GRAVEL

The river terrace deposits (Chapter 8) are a resource of gravel, but have not been widely exploited. None of the deposits has been penetrated by boreholes, so that thicknesses and quality are uncertain; thickness estimates of up to 34 m, from resistivity soundings (p.108) on the extensive low level gravel spreads between Porlock and Bossington and south of Allerford, are likely to be excessive. The higher-level river terrace deposits probably do not normally exceed 3 m in thickness. The only recorded working in the river terrace deposits was from a now mainly infilled pit [SS 999 447], north-east of Marsh Street, where the gravels were at least 2 m thick.

Weathered Otter Sandstone represents a resource of sand, but most workings in the formation were probably for building stone (see below).

There is no known offshore dredging of sand and gravel in the district, the main dredging areas being farther east in the Bristol Channel (Tappin et al., 1994, fig. 69).

ROADSTONE

Many small disused quarries and pits in the Hangman Sandstone and in head were probably worked for roadstone as well as building stone. The Hangman Sandstone remains a large resource of roadstone, but most of the outcrop lies within Exmoor National Park and planning policies, as outlined in the Minerals Local Plan, make it most unlikely that any proposals for large-scale working would be approved. Larger quarries in the Hangman Sandstone, for example Conygar Quarry [SS 9925 4413], Conygar Wood Quarry [SS 9893 4426] and Ellenborough Quarry [SS 9875 4423], all near Dunster, were probably worked for roadstone, but no quarries are open at present. No details of Polished Stone Values (PSV) or Aggregate Abrasion Values (AAV) are available for the Hangman Sandstone of the district. The nearest comparable locality for which figures are available are from the Hangman Sandstone in Triscombe Quarry [ST 161 356] in the Quantock Hills. The figures show generally high PSV and low AAV which, taken together with the consistent Aggregate Impact Value and Aggregate Crushing Value data, confirm the locality to be a source of high-quality gritstone (Cox et al., 1986).

### BUILDING STONE

#### Devonian

Many small quarries scattered over the Hangman Sandstone outcrop testify to its use in buildings, usually near the site of the quarry. There is an abundant supply of rubbly surface stone from the Hangman Sandstone over much of the district, but it is hard and difficult to dress and was used in ragwork or rubble fashion in many buildings. Ancient churches (e.g. at Culbone), have rubble towers, but some (e.g. at Watchet) have towers of massive ashlar blocks.

#### Permo-Triassic

The main sources of Permo-Triassic building stone were the better-cemented sandstones of the Otter Sandstone; the Luccombe Breccia has also been used as a building stone, as have sandstones from the basal part of the Mercia Mudstone in the Porlock Basin.

Quarries scattered over the Luccombe Breccia outcrop supplied building stone, although some were also worked for lime (see below). Buildings on the outcrop are commonly built of the local breccia, for example Holt [SS 9236 4480] and Gillhams [SS 9211 4452].

Warm, pale reddish brown (locally greyish green) Otter Sandstone has been widely used in buildings in Minehead town. The source of much of the sandstone was probably Staunton Quarry [SS 973 449], Alcombe, although there were also quarries in the town, for example that recorded by Ussher (BGS, MS) near St Michael's Church, Higher Town (exact location uncertain), as having been a source of building stone. The Budleigh Salterton Pebble Beds at Alcombe have been used in walls and buildings in the vicinity of the quarry.

Burrowhayes Quarry [SS 8985 4613], West Luccombe, in basal or marginal calcareous sandstones of the Mercia Mudstone, is reported to have produced building stone which was supplied to Porlock and Minehead in the 1930s, but found to be too porous for domestic buildings.

#### Jurassic

Lias Group limestones have been used for buildings in the Selworthy and Watchet areas. In Minehead, the tower of St Michael's Church, Higher Town, is mainly of Lias limestone with some Triassic sandstone. Old pits [SS 916 463 to 926 461] in the basal beds of the Blue Lias south of Selworthy, between Holnicote and East Lynch, were probably mainly for marl, but may have yielded some building stone. In the adjacent district, Whittaker and Green (1983, p.117) noted that the calcareous beds of the lowermost 13 m of the 'Lower Lias' were most commonly worked, and many beds were given names appropriate to their uses or properties, e.g. 'Paviour', 'Building Stone' and 'Hearth Stone'. The limestones were also used as paving stones and walling stones.

Some imported Jurassic stones, such as Ham Hill Stone, have been used in buildings in Minehead (Prudden and Edwards, 1994).

#### Drift

Large split cobbles of Hangman Sandstone, probably from the shingle ridge, have been used in some buildings in Minehead, for example in buildings close to the junction of North Road and The Avenue (Prudden and Edwards, 1994).

Head deposits probably provided a source of rubbly sandstone for use in the construction of field walls and farm buildings.

### BRICK AND TILE CLAY

The main resource of brick and tile clay in the district is the Mercia Mudstone, which was formerly worked in a small number of pits; there are now no active workings. The largest pit was that at Alcombe [SS 972 453], now a caravan site. The kilns, now demolished, were on the north side of the main road, and the clays were transported over the road by aerial ropeway. The buildings are marked on the 1948 edition of the 1:10 560 scale Ordnance Survey Map as 'Victoria Brick and Tile Works'. Another pit [ST 032 435] at Blue Anchor was apparently worked for brick clay, as suggested by the name 'Brickyard Covert' nearby on the 1928 1:10 560 scale Ordnance Survey map. Most other pits on the Mercia Mudstone outcrop were probably worked for marl (see below).

### MARL AND LIME

Although the Luccombe Breccia is dominantly composed of Devonian sandstone and slate clasts, Horner (1816) noted that thick veins of calcite had locally been worked for lime, as for example in Gillhams Quarry [SS 9197 4445], about 1 km east of Luccombe (p.61).

The Budleigh Salterton Pebble Beds conglomerate at Staunton Quarry [SS 973 449] and Alcombe Quarry [SS 976 448], south of Minehead, was formerly worked for lime. Pebbles and cobbles of Carboniferous Limestone were picked out from the conglomerate for this purpose and burnt in nearby kilns (Horner, 1816).

Small pits on the Mercia Mudstone outcrop, of which there are several between Wydon Farm and Woodcombe, were probably dug to obtain marl for spreading on the land for agricultural purposes.

Old pits [SS 916 463 to 926 461] in the basal beds of the Blue Lias south of Selworthy, between Holnicote and East Lynch, were probably dug mainly for marl.

GYPSUM AND ALABASTER

Veins and nodular masses of colourless, white and red-tinged gypsum are present throughout the 'Grey Marl' division of the Blue Anchor Formation between Watchet and Blue Anchor, but are abundant only in beds below 10.67 m from the top of the formation, particularly in two units (A and B, of Warrington and Whittaker, 1984) (Figure 26). Working was taking place at the time of De la Beche, who noted (1839) that the gypsum was sent to various places along the Bristol Channel. Up to eight beds varying from about 0.2 to 1.7 m thick were dug in the Warren Rocks Gypsum Workings until about 1923 (Sherlock and Hollingworth, 1938). The beds were irregular and laterally impersistent, with much gypsum in the form of veins and strings connecting the larger masses. The gypsum was dug out of the face of the cliff in ledges which were only a few metres deep owing to the steepness of the cliff face. Levels were not driven into the cliff because of the irregularity of the deposits. The gypsum was thrown down onto the foreshore and carted along the foot of the cliff to Watchet, except when the workings were near Blue Anchor, in which case it was sent to Watchet Harbour by boat. Sherlock and Hollingworth (1938) noted that the gypsum was used as a fertiliser at various places along the Bristol Channel, and also for paper-filling.

Alabaster is a (usually) fine-grained and translucent variety of gypsum capable of being carved and used as an internal decorative stone. Firman (1984, 1989) has described the seventeenth century Somerset industry, centred on Watchet, which supplied alabaster to much of south-west England, mainly for use in church monuments.

CELESTINE

Whittaker and Green (1983) noted that small deposits of celestine (celestite: strontium sulphate) occurred locally at or near the base of the Blue Anchor Formation (Mercia Mudstone Group) and may have been dug on a very small scale in the past (Nickless et al., 1976).

## Severn Barrage

Various schemes have been proposed since 1849 to build a barrage across the Severn estuary (Clarke, 1982; Kirby, 1988; Tappin et al., 1994). Following a recommendation from the Select Committee on Science and Technology in 1977, a report published in 1981 identified two prospective sites for a barrage. One lay partly in this district, following an approximately north–south line from Warren Point east of Minehead to Breaksea Point in South Wales. The second, preferred, location lay farther east, between Lavernock Point and Brean Down. The barrage sites were chosen particularly to follow areas where there was little sediment cover, in order to reduce costs of foundation construction. The scheme was considered to be technically

feasible and to have a generating capacity of 7200 MW with an annual energy output of 12.9 TWh, at a cost of £5660 million (1981 prices). A further study (Clare, 1987) confirmed much of the earlier work.

## Hydrocarbons

The Mesozoic hydrocarbon potential of the Bristol Channel Basin has been discussed by Kamerling (1979) and Nemčok et al. (1995). No hydrocarbon shows have so far been reported from wells drilled in the western part of the basin, owing, firstly, to insufficient hydrocarbon generation from the patchily developed, rather low-quality Lias Group source rocks, and, secondly, to the lack of suitable traps existing at the time of hydrocarbon generation. In addition, there is an almost complete lack of suitable reservoir rocks. Near-mature Lias Group crops out in South Wales and Somerset (Cornford, 1986). At Kilve, east of the district, a 'Lower Lias' bituminous shale was worked as a source of oil, yielding 40 gallons per ton. The possible presence of Silesian rocks beneath the offshore area (Chapter 3) could offer older source rocks. Maturities are variable (Smith, 1993). South of the Exmoor–Cannington Park Thrust, the rocks are over-mature, but between that thrust and South Wales, Silesian rocks are likely to lie within the gas window, although the effects of Mesozoic burial, which was substantial, are not known from actual measurements (Maddox et al., 1995). The inner part of the Bristol Channel was not considered favourable for hydrocarbons, but the possibility that Silesian rocks are present, combined with the existence of a reservoir (e.g., Triassic sandstones), is significant for the hydrocarbon prospectivity of the area.

## Mining

Metalliferous mining was not important in the district, and the only known workings were for iron in mineralised Luccombe Breccia at Wychanger or Knowle Top Mine [SS 913 445] near Luccombe (Dines, 1956). This mineralisation is described in Chapter 5. The site is marked by two open workings trending approximately west-north-west to east-south-east. Within these trenches, two ponds [SS 9132 4455, SS 9131 4447] mark the sites of deeper excavations, probably shallow shafts. All traces of a drainage adit, reported to have been situated to the west [SS 9122 4457] of the northern open working, have been obliterated. There are no known plans of the workings and the possibility of underground workings of unknown extent cannot be excluded, and should be taken into account in the event of any development in the area. The deposit was apparently first noted by De la Beche (1839, p.617) who recorded that the iron ore was shipped to South Wales from Porlock and Minehead. There is no evidence of other workings 'a quarter of a mile NW of Luccombe' (possibly [SS 9099 4484]) referred to by Dines (1956).

Beer (1988) reported that the mine was apparently worked together with Brockwell Mine, south of the district. They were producing in 1837, but had ceased by 1858. They reopened for a short time in 1870.

Horner (1816, p.349) noted a shaft sunk for copper in 'one of the branches of Grabbist Hill at Staunton, above the village of Alcombe', but the quantity of ore found was so small that the working was very soon abandoned. This locality was not found during the survey. The same mine was mentioned by De la Beche (1839). Beer (1988) noted that ore from this mine was exhibited, with that from nearby properties at Alcombe, North Hill and Dunkery, in 1855; the precise location of these properties is unknown. The mine at Alcombe was reputedly restarted in 1870 and found some ore, but was closed soon afterwards by litigation. In 1870, trials for iron ore were begun at Hopcott 'presumably on the Common to the S.W. of Minehead' (Beer, 1988).

The purpose for which Gillhams Quarry [SS 9197 4445] was opened is not clear; building stone may have been worked from the harder beds of breccia and sandstone, and material from the calcite vein complex described in Chapter 5 could have been worked for lime burning. The form of the western quarry suggests that it may have been a mining trial, possibly for iron.

## RADON

Radon is a naturally occurring radioactive gas produced by the decay of both uranium and thorium which are present in all soils and rocks. The gas can be a potential health hazard in buildings if high levels accumulate, and it has been estimated by the National Radiological Protection Board (NRPB) that at least half of the total radiation dose for the average Briton is obtained from combined radon and thoron (Clarke and Southwood, 1989). There is a direct link between geology and the levels of radon generated at the surface.

Radon is easily dispersed to very low levels in the atmosphere, However, in confined spaces in contact with soil or rock, such as caves, mines and buildings, radon and its decay products can accumulate. Soil gas is drawn into a building by a slight underpressure indoors which results from the warmer air rising. Hence, radon problems in houses are due to bulk flow of ground gas carrying radon, with a relatively small contribution by the diffusion of radon through and from building materials. The amount of ground air drawn into a house will vary according to the local permeability of the ground and the nature of the leakage into the house.

In January 1990, the NRPB issued revised advice on radon in homes, resulting in a new Action Level for radon of 200 Becquerels per cubic metre ($Bq/m^3$). Parts of the country in which there were one per cent or more of the homes above the Action Level were to be regarded as 'Affected Areas'. The NRPB recommended that within Affected Areas, the government should designate localities where precautions against radon in new houses were required. A classification of areas requiring some form of building control was adopted by the Building Research Establishment in 1996. If more than 10 per cent of existing houses are affected (high risk), all new dwellings in the area should incorporate full radon protection (i.e. a radon-proof barrier), along with provision for sub-floor

ventilation measures should barriers prove ineffective. Where between 3 and 10 per cent of the houses are affected (moderate risk), all new dwellings should have provision for future sub-floor ventilation if required. In such a case, the house would be built complete with sump and extract pipe which could be connected to a fan if measurement revealed that there was a problem with radon ingress.

Areas where high levels of radon may exist in dwellings can be recognised using two approaches: (i) by measuring the levels of the gas in individual houses, or (ii) by estimating the radon emission potential of the ground using measurements of radon concentrations in soil gas combined with geological and other factors that influence the emission of the gas at the surface, such as permeability and uranium concentration. A study of Somerset (Appleton and Ball, 1995) showed that there is generally good agreement between the radon potential class (based on soil gas radon and permeability) and the house radon risk class (based on house radon measurements) for most of the geological units. Some of the divergences are attributed to insufficient data. The resulting Radon Potential Map was based on a classification of geological units using all available soil gas radon, permeability and house radon data.

The geological units in the district fall into the Low or Moderate Radon Potential classes. The Moderate Class is equivalent to the Secondary Protective Measures zone of the Buildings Research Establishment Protective Measures Map, in which 3 to 10 per cent of the houses are above the Action Level of 200 $Bq/m^3$. The Low Class is equivalent to a zone in which less than 3 per cent of houses exceed the Action Level. Units within the district having Moderate Radon Potential are: Hangman Sandstone, Penarth Group, Blue Lias, river terrace deposits, alluvium, and head. The following geological units have Low Radon Potential: Mercia Mudstone Group, peat, blown sand, alluvial cone, and marine deposits.

Radon potential mapping cannot be used to predict whether an individual existing or new house will be affected, since this depends on factors such as construction methods, ventilation, and how the house is used. However, it can be used to delineate potentially affected areas and should be more precise than using house radon data averaged over 5 km squares, as is done on the NRPB house radon maps.

Householders concerned about the levels of radon in their property are advised to contact the NRPB who will advise on whether measurements for levels of radon should be carried out.

## LAND USE, SOILS AND AGRICULTURE

The district is mainly rural, the only substantial built-up areas being those centred on Minehead, Watchet and Porlock. The main development during the last 15 years in the Minehead area has been on the east side of the town. Also east of the town, a large holiday camp [SS 985 460] has been constructed on made ground overlying saltmarsh deposits. A small industrial estate is present east of the town, and also is partly developed on saltmarsh deposits.

The main land use outside the built-up areas is for agriculture, with some areas of woodland. The coastal slope is wooded west of Porlock Weir, and there are also woods between Bossington and Selworthy, on the south side of Selworthy Beacon, around Horner, Luccombe and along Hawk Combe, and between Tivington and Dunster. Horner Wood is an area of ancient oak woodland noted for its lichen flora. Areas of upland moor are present between Bossington Hill and North Hill, west of Minehead, and south and west of Porlock.

The agricultural land classification is shown on Sheet 164 of the Ministry of Agriculture 1:63 360 scale map (1971). The lowlands underlain by Permo-Triassic rocks — the Porlock Basin and the Minehead Basin — contain the best quality farmland in the district, with substantial areas of grade 2 as well as grade 3 land; one area of about 1 km² of grade 1 land is present in the area centred on Blackford. The higher ground mainly underlain by the Hangman Sandstone is mostly grade 4 and 5 land.

The distribution of soil associations in the district is shown on the 1:250 000 scale Soil Survey map 'Soils of South West England'. The soils and their use are described in Findlay et al. (1984), from which the following account is summarised.

### Soils on Devonian rocks

Soils on the higher hilltops in the western half of the district are very to extremely acid ferric podzols and stagno-podzols of the Larkbarrow association, developed on blanket head and regolith overlying Hangman Sandstone. These soils have a thin (mostly less than 0.15 m) black surface layer of peat or humose, sandy, silt loam, overlying a bleached, stony, sandy, silt loam subsurface layer. An ironpan layer is locally present at the base of the leached horizon. On valley sides, and at lower altitudes on the outcrop of Devonian rocks, an association (Rivington 2) of mainly well-drained, loamy brown earths and brown podzolic soils is developed.

Much of the higher ground occupied by Larkbarrow soils is heather moor, but gentle slopes at lower levels have been reclaimed for pasture and occasional forage crops, especially near the coast. Rivington 2 soils in the district are largely under ley or permanent pasture, with some semi-natural vegetation, and fairly extensive areas of coastal and inland woodland.

### Soils on Permo-Triassic rocks

Breccias between Luccombe and Huntscott carry mainly well-drained, gravelly, reddish brown loams of the Crediton association. Between Allerford and Luccombe the soils are reddish brown, fine loamy brown earths of the Milford association. The Mercia Mudstone outcrop is characterised by reddish brown clayey soils of the Worcester association, generally with a thin (about 0.3 m) surface layer of clay loam containing pebbles of Devonian sandstone.

Much of the land characterised by Crediton and Milford soils is used for mixed arable and grassland. The Worcester soils are mostly under permanent grassland, but some winter cereals are grown.

### Soils on Jurassic rocks

Dark greyish brown to olive-brown calcareous clay soils of the Evesham 2 association are developed on the Lias Group around Selworthy and between Blue Anchor and Watchet. They are slowly permeable, and thus waterlogged for long periods in winter. The soils are mostly under grass.

### Soils on Quaternary deposits

Soils developed on marine deposits behind the shingle ridge of Porlock Bay, and north of Alcombe, are non-calcareous and clayey (Wallasea 1 association). The shingle ridge in Porlock Bay is liable to be breached and flood the saltmarsh behind it. The land is under permanent grass. Spreads of alluvial gravel landward of the coastal marsh deposits in Porlock Vale and around Marsh Street and Blue Anchor are associated with reddish brown, mainly well-drained, loam over gravel at depths which may be 0.4 m or greater (Newnham association). They are versatile soils allowing cultivation of a wide range of crops. Farther east in Porlock Vale, between Holnicote and Blackford, less well-drained soils (Wigton Moor association) are developed in river terrace deposits.

## CONSERVATION

The special landscape value of much of the area within the district west of Minehead was recognised in 1954 by the designation of the Exmoor National Park. The conservation policies affecting the park are set out in the Local Plan. The National Trust, which owns the Holnicote Estate between Porlock and Minehead, is also committed to protection from unsuitable development of the land that it owns. The coast west of Minehead is defined as 'Heritage Coast'.

Additional protection is given to Sites of Special Scientific Interest (SSSIs) which are areas worthy of conservation and protection because of their flora, fauna, geological or physiographical features. The existence of a site is relevant to planning and development decisions. Sites are notified to the Department of the Environment, local authorities, and to the owners of the sites. Notifications were made under section 23 of the National Parks and Access to Countryside Act, 1949, or section 28 of the Wildlife and Countryside Act 1981. The district contains five SSSIs, the boundaries of which are shown on Figure 3. Two of the sites (Glenthorne; and Blue Anchor to Lilstock Coast) are coastal localities which were notified primarily for their geological importance; the remainder are of interest for their fauna, flora and habitats, and comprise Porlock Marsh, Exmoor Coastal Heaths and North Exmoor.

In 1995, English Nature designated 4000 acres of Horner Wood and Dunkery as a National Nature Reserve.

*Glenthorne SSSI.* This geological SSSI extends from [SS 794 499] near Giant's Rib to [SS 805 495] near The Caves (Figure 3). It was notified under the 1981 Act in 1989 as an accessible and well-exposed section in the Hangman Sandstone.

*Blue Anchor to Lilstock Coast SSSI.* This geological SSSI [ST 033 435 to ST 195 462] lies partly within the district (Figure 3). It was notified under the 1949 Act in 1971, and under the 1981 Act in 1986, owing mainly to the outstanding series of sections through Late Triassic and Early Jurassic rocks, which are of international importance. Cliffs in St Audrie's Bay [ST 1020 4330], about 3 km east of the district boundary, have been proposed as a possible Global Stratotype Section and Point for the base of the Hettangian Stage, and thus of the Jurassic System (Warrington et al., 1994). Blue Anchor Cliff [ST 0385 4368] is the type section of the Blue Anchor Formation at the top of the Mercia Mudstone Group (Warrington and Whittaker, 1984), and in this area also the Penarth Group is well exposed. The area is also important for coastal geomorphology, and shows well-developed intertidal shore platforms, locally veneered by mud, sand and shingle, and reflecting in detail the variable resistance to erosion of the Mercia Mudstone, Penarth Group and Blue Lias. The platforms developed in a macrotidal environment and are among the best examples of such coastal features in Britain.

*Porlock Marsh SSSI* [SS 880 479]. The site (Figure 3) was notified under the 1981 Act in 1990, owing to its importance for strandline, shingle, maritime grassland, saltmarsh, swamp and brackish-water ditch habitats. Porlock Marsh is sited behind the striking shingle ridge that extends across Porlock Bay (Plate 27). It is underlain by Quaternary deposits (storm beach and saltmarsh deposits).

*North Exmoor SSSI.* Part of the eastern portion of the SSSI lies within the district between Oare and Horner (Figure 3) and is underlain by Hangman Sandstone. The site was notified under the 1949 Act in 1954, and under the 1981 Act in 1992. It is nationally important for its south-western lowland heath communities, and for transitions from ancient seminatural woodland through upland heath to blanket mire (outside the district). The ancient woodland is mostly found around Horner and Hawk Combe, within the district, and is nationally important for its lichen fauna, over 320 species being represented.

*Exmoor Coastal Heaths SSSI* [SS 920 480]. This SSSI has only recently been notified (1995) and was unconfirmed at the time of writing. That part of the SSSI lying within the district extends along North Hill from Hurlstone Point to near Minehead (Figure 3) and was notified under the 1981 Act in 1995. It is underlain by Hangman Sandstone. Its importance lies in the extensive areas of heathland communities represented which are rare in Britain, or confined largely to South-west England and South Wales. There are also important upland heath, mire and grassland habitats and, at lower altitudes and in the coastal zone, woodland and scrub, and acidic and maritime grassland. Nationally rare and scarce plants are associated especially with the coastal communities and woodland, and a breeding colony of a nationally rare butterfly also occurs.

The Exmoor National Park Local Plan identifies County Geological Sites of local conservation importance which are evaluated using the same criteria as those for Regionally Important Geological/Geomorphological Sites (RIGS). In the present district, land between Porlock Weir and Worthy, and at Greenaleigh has been identified as County Geological Sites (Figure 3). A list of RIGS, together with boundary and citation informations is available at the Somerset Environmental Records Centre (p.115).

# THREE

# Concealed geology

## INTRODUCTION AND SUMMARY OF CONCEALED GEOLOGY

Information about the concealed geology of the district has, in the absence of boreholes penetrating to the concealed strata, been obtained wholly from a study of geophysical data. Of particular significance in evaluating the offshore area are five commercial multichannel seismic reflection profiles which revealed important details of crustal structure to depths of about 5 km. The seismic sections illustrated and described in this chapter come from a variety of non-exclusive proprietary surveys owned by Geco-Prakla, a division of Geco Geophysical Company Ltd. Regional gravity and aeromagnetic surveys for the district provide further data on deep structure. An onshore gravity survey carried out to investigate the form of the Porlock Basin gave ambiguous results, owing to the lack of density contrast between the Luccombe Breccia (Permo-Triassic) and the underlying Devonian rocks.

The seismic reflection profiles show that the offshore area is occupied by a synclinal basin of Mesozoic rocks (the Bristol Channel Syncline), containing Triassic and Lower to Upper Jurassic strata, and is crossed by the east–west-trending Central Bristol Channel Fault Zone. The interpretation of the deeper offshore geology is more problematical, but it is here considered possible that the Mesozoic rocks are underlain mostly by Silesian rocks which rest, probably conformably, on Carboniferous Limestone. In the southern part of the district, however, Devonian rocks are thrust over a high velocity (6.2 km s$^{-1}$) unit which underlies a prominent seismic reflector (Reflector X, of Brooks at al., 1993). The high-velocity unit was considered, by Brooks et al. (1993), to comprise pre-Devonian schistose or gneissose basement rocks. However, there is a possibility that it could comprise Carboniferous Limestone, and that Devonian rocks could, therefore, be thrust northwards over Carboniferous rocks on a westward extension of the Exmoor–Cannington Park Thrust, first recognised near Cannington Park, about 25 km east of the district. This interpretation is conjectural, owing to difficulty in deciding on the nature of the high-velocity unit.

Other evidence from seismic reflection profiles (Brooks et al., 1988) indicates that Reflector X lies in the hanging-wall block of a major Variscan thrust (the Bristol Channel Thrust) which dips south at an average of 25° (Brooks et al., 1993). Reactivation in extension of this thrust is believed to have resulted in the formation of the Central Bristol Channel Fault Zone (Brooks et al., 1988).

The Bristol Channel Thrust is not well imaged on the seismic profiles within the district. Pre-Carboniferous reflectors can be identified on the offshore seismic profiles and indicate the probable presence of Lower Cambrian rocks. Ordovician rocks (particularly Tremadoc strata) are also likely to be present, but have not been definitely identified. Silurian rocks have not been recognised, but may be present in the east of the district. Seismic refraction data indicate that Precambrian crystalline rocks underlie the Bristol Channel Basin at depths of 5.5 to 6.5 km, and are at a depth of about 7.5 km beneath the north Devon coast (Mechie and Brooks, 1984).

## POTENTIAL FIELD GEOPHYSICAL DATA

Regional geophysical data for the district include the results of gravity and aeromagnetic surveys, interpretation of which relies on a knowledge of the physical properties of the main rock units.

### Gravity data

Bouguer anomaly data for the district were first acquired and interpreted by Bott et al. (1958), as part of their regional study of South-west England. Later surveys were carried out by the BGS using a greater station density. Brooks and Thompson (1973) provided the initial offshore coverage for part of the Bristol Channel; this was also augmented by later BGS surveys. The onshore and offshore BGS gravity data were used to generate the regional Bouguer anomaly map (Figure 4). The southern part of the map (including all the land area) is dominated by an extensive linear zone of northward decreasing Bouguer values (the Exmoor Gradient Zone); the district lies at the northern margin of this zone. Bouguer values reach a minimum in the Bristol Channel over the thickest part of the Mesozoic basin, in the western offshore part of the district (Figure 4). Superimposed on the Exmoor Gradient Zone are lower amplitude anomalies due to near-surface density changes; these are discussed below.

Additional gravity data were collected in the Vale of Porlock during the course of the survey with the aim of modelling the form of the post-Variscan basin there. These new data include two detailed traverses across the basin, and infill of the regional coverage.

**Figure 4**  Bouguer gravity anomaly map and simplified geology of the district and adjacent areas. Contour intervals at 1 mGal onshore and 2 mGal offshore. Line AA′–location of Bouguer gravity profile, see Figure 7.

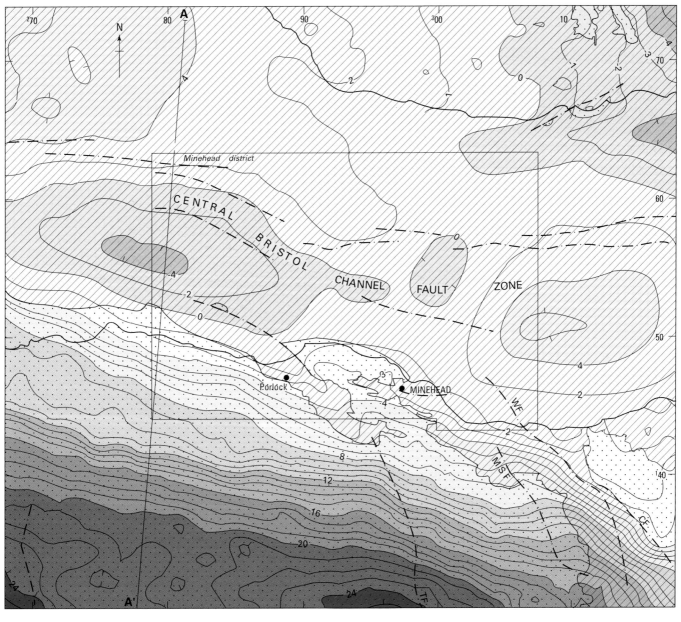

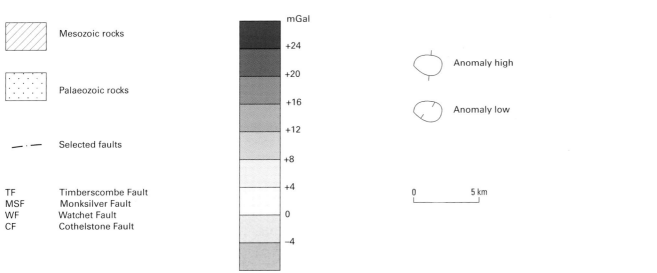

Mesozoic rocks

Palaeozoic rocks

Selected faults

TF     Timberscombe Fault
MSF    Monksilver Fault
WF     Watchet Fault
CF     Cothelstone Fault

mGal

+24

+20

+16

+12

+8

+4

0

−4

Anomaly high

Anomaly low

0       5 km

## Aeromagnetic data

Aeromagnetic data for the district were acquired in the following three surveys, carried out for the BGS:

1:  onshore, flown at 152 m mean terrain clearance with north–south flight lines 0.4 km apart, and east–west tie lines 10 km apart.

2:  offshore, east of National Grid line [300E], and onshore east of National Grid line [310E], flown at 549 m barometric height with north–south flight lines 2 km apart, and east–west tie lines 10 km apart.

3:  offshore, west of National Grid line [300E], flown at 305 m mean terrain clearance with north–south flight lines 2 km apart, and east–west tie lines 10 km apart.

The surveys were carried out with some overlap in the area between [300E] and [310E] onshore and [398E] and [302E] offshore. The data from the three surveys have been adjusted to produce Figure 5.

The total field magnetic anomalies relative to a computed regional field for the British Isles are shown in Figure 5; two sets of anomalies with contrasting wavelengths are apparent. The most obvious short wavelength anomalies are two linear features which trend west-north-west–east-south-east across Exmoor and have a near-surface origin. Longer wavelength anomalies, reflecting the presence of deep-seated magnetic basement, show a gradual increase of values to the east and, in the northern part of the area shown in Figure 5, to the north.

## Physical properties

The densities of the main rock units are summarised in Table 3. There is little contrast between the Mesozoic and older rocks, due largely to the relatively high densities of the Permian(?) and Triassic rocks. Within the Devonian, arenaceous and argillaceous units are characterised by significantly different densities: in the Minehead district this is reflected by a contrast of about 0.10 $Mgm^{-3}$ between the Hangman Sandstone and the Lynton Formation.

Density and porosity measurements were also made on 15 samples of Devonian and Triassic rocks collected during the present mapping (Table 4). The density of the Hangman Sandstone samples is slightly higher than that reported previously (Table 3) and is not significantly different from that of the Luccombe Breccia. Although the mean densities of these rocks are similar, the Luccombe Breccia has a higher porosity (Table 4) which is compensated for by a higher grain density, reflecting the presence of mudstone/slate fragments. The two samples of Otter Sandstone have contrasting saturated densities of 2.04 and 2.56 $Mgm^{-3}$, the lower value probably indicating the removal of most of the cement.

Most sedimentary rocks have low magnetic susceptibilities and are therefore unlikely to give rise to significant magnetic anomalies. However, pyrrhotite-bearing slates in the Honeymead boreholes [SS 7989 3935] (Edmonds et al., 1985), just to the south of the district, have an aver-age susceptibility value of $0.72 \times 10^{-3}$ (SI units), with maximum values of up to $10 \times 10^{-3}$. The rocks also have a strong remanent magnetisation and this is largely responsible for the pronounced west-north-west-trending linear magnetic anomalies (Figure 5) (Edmonds et al., 1985; Jones et al., 1987; Thomson et al., 1991).

## INTERPRETATION OF POTENTIAL FIELD GEOPHYSICAL DATA

Both the Bouguer gravity and the aeromagnetic data indicate anomalies of different wavelengths which are suggestive of sources at different depths. Digital filtering techniques were applied in an attempt to isolate the components of the two fields; for example, polynomial residual anomalies were calculated to isolate Bouguer anomalies of near-surface origin (Figures 6, 9).

Detailed interpretations were carried out along selected profiles using a 2.5D modelling program. The most important profile relevant to the deep structure is the north-south line AA′ (Figures 4, 7, 8), which includes the main Exmoor Gradient Zone.

The profile (Figure 7) includes the seismic profile B (Figures 11,12) and has been interpreted using constraints imposed by the seismic evidence, including the locations of Reflector X (Brooks et al., 1993) and the high-velocity refractors, and other geological evidence. With these constraints the modelling suggests that the main cause of the Exmoor Gradient Zone does not lie within the upper few kilometres of the crust but is a more deep-seated density change, probably associated with the underlying Lower Palaeozoic and Precambrian rocks. This change is considered mainly to reflect the abutment of two basement types along a major structure related to the thrusting seen in the upper part of the crust. The southward thinning of the Devonian rocks illustrated in the model is preferred on geological grounds, but cannot be constrained closely by the gravity evidence; increasing the thickness of the lower-density Devonian rocks could be compensated for in the calculated profile by increasing the thickness or density of the high-density basement. Although the Mesozoic rocks in the Bristol Channel Basin produce a gravity low, its amplitude is not very great because of the relatively high density of these rocks. It was necessary in the model (Figure 7) to introduce a wedge of low-density (2.60 $Mgm^{-3}$) material beneath the southern part of the channel which could represent a sequence of Devonian rocks thickened along the major deep structure.

It can be seen that, due to its location near the minimum gravity anomaly, the seismically defined thrust block (TB) needs to be of relatively low density, in addition to having high velocity and low magnetisation. In the interpretation shown in Figure 7, the density (2.70 $Mgm^{-3}$) ascribed to the lower part (below Reflector X) of this block is consistent with the Lower Carboniferous limestone sequence indicated by the seismic interpretation presented in this memoir. This unit has been continued southwards beneath the Devon coast to form the high-velocity refractor, although interpretation of the gravity profile

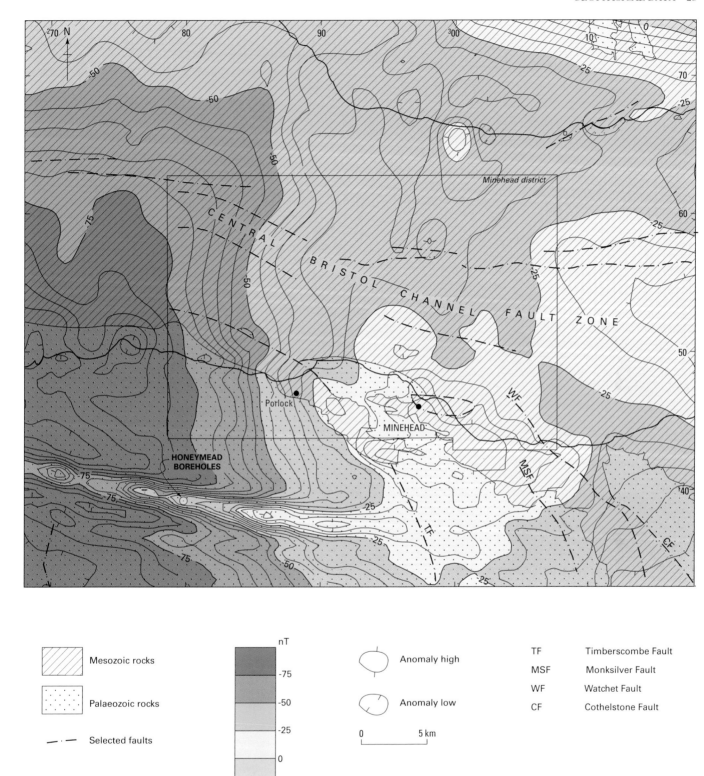

**Figure 5** Aeromagnetic anomaly map of the district and adjacent areas, with contours at 5 nT intervals. The map is based on adjusted data from three surveys with different flight heights and line separations (see text).

**Table 3**  Typical densities of the main lithostratigraphical units in the Bristol Channel area, based on Bott et al. (1958), Al-Saadi (1967), Whittaker and Scrivener (1982), Whittaker and Green (1983) and Edmonds and Williams (1985).

| Rock unit | Density (Mgm$^{-3}$) |
|---|---|
| **Jurassic** | |
| Lias Group | 2.54 |
| **Triassic** | |
| Penarth Group | 2.46 |
| Mercia Mudstone Group | 2.64 |
| Sherwood Sandstone Group | 2.68 |
| **Carboniferous** | |
| Coal Measures | 2.66 |
| Pennant Measures | 2.62 |
| Lower Carboniferous | 2.65 |
| Culm | 2.69 |
| **Devonian** | |
| Ilfracombe Slates | 2.72 |
| Hangman Sandstone Formation | 2.55 |
| Lynton Formation | 2.65 |

would be improved by the presence of lower-density Devonian rocks beneath the Dinantian limestones which are estimated to have a thickness of about 1.5 km.

Also shown on the profile (Figure 7) are the pyrrhotite-bearing rocks (P) of Exmoor.

**Lineaments**

Geophysical lineaments, as defined on the basis of gravity and aeromagnetic data, usually have a structural significance rather than being a manifestation of lithological variations. In the area shown on Figure 8, centred on the Minehead district, the main lineaments are:

A. a dominant group with east-south-east trends, paralleling the strike of the Devonian strata (e.g. G1, G2, M1, M2)

B. a group with south-east trends (T1, T2).

It is possible that the appearances of some of the gravity lineaments in group A are enhanced by their coincidence with strike faults, but this cannot be demonstrated without more detailed gravity data. Magnetic lineaments

M1 and M2 (Figure 8) correspond to the pronounced linear Exmoor magnetic anomalies and also apparently follow the strike of the Devonian rocks just south of the district. These anomalies appear to form part of a more extensive geophysical lineament that extends eastwards for at least 120 km from the coast near Ilfracombe through the Quantock Hills to the Somerton Anticline (Cornwell, 1986).

Group B lineaments have trends similar to those of transcurrent faults in north Devon and west Somerset. Several features in this group are indicated (Figure 8); the main ones in the district are:

**T1**  defined largely on the basis of truncation or dextral displacement of lineaments G2, M1 and M2, and apparently associated there with the Timberscombe Fault. The lineament continues north-westwards close to the north-east margin of the Porlock Basin.

**T2**  the Watchet–Cothelstone fault system is not well defined geophysically onshore, but the gravity feature T2, and a weak aeromagnetic feature, probably represent its extension into the Bristol Channel. South-east-trending faults close to lineament T2 have been identified on seismic profiles D and G (Figure 11), just north of the Central Bristol Channel Fault Zone, and may represent the offshore extension of the Watchet–Cothelstone fault system.

East of lineament T2, a gravity high centred on [ST 09 50] (Figure 4) suggests that high-density basement underlies Mesozoic rocks in Bridgwater Bay (Brooks and Thompson, 1973; Whittaker and Green, 1983, p.115).

**Structure in Devonian and older rocks**

The Exmoor Gradient Zone included in the profile AA′ (Figures 4, 7) has been the subject of several previous investigations, mainly because of its possible significance for the existence of large-scale thrusting in north Devon. The gravity contours parallel the geological strike, but the surface geology provides no obvious explanation for the anomaly. Falcon (in Cook and Thirlaway, 1952) suggested that a continuation of the gradient zone seen in the Quantock Hills could reflect the presence of a thrust postulated, on geological evidence, farther east in the Cannington Park area. Bott et al. (1958) subsequently discovered the gradient zone over Exmoor and suggested that it was due to a wedge of Devonian rocks, thinning

**Table 4**  Densities and porosities of some Devonian and Permo-Triassic rocks. The mean values and standard deviations (in brackets) and number of samples (N) are listed.

| Unit | N | Densities (Mgm$^{-3}$) | | Porosity (%) |
|---|---|---|---|---|
| | | Saturated | Grain | |
| **Permo-Triassic** | | | | |
| Otter Sandstone | 2 | 2.30 | 2.67 | 22.2 |
| Luccombe Breccia | 9 | 2.61 (0.03) | 2.72 (0.03) | 6.7 (1.4) |
| **Devonian** | | | | |
| Hangman Sandstone | 4 | 2.62 (0.04) | 2.64 (0.03) | 1.2 (0.7) |

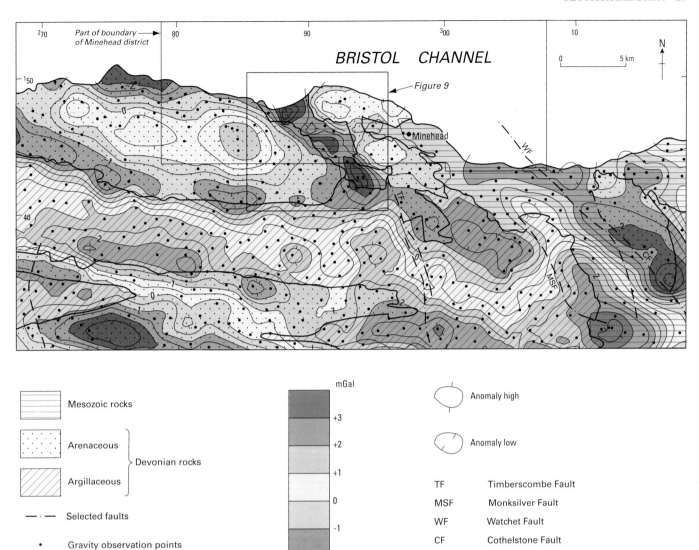

**Figure 6** Residual Bouguer gravity anomaly map for the onshore part of the district, and adjacent onshore areas, with gravity observation points and simplified geology. The residual anomalies, with contours at 0.5 mGal intervals, were derived by removing a 3rd order polynomial regional field from the observed data (Figure 4).

northwards and thrust over lower-density Devonian or Carboniferous rocks; other explanations considered included a northward thickening of Lower Devonian rocks.

The acquisition of offshore gravity data for the Bristol Channel area (Brooks and Thompson, 1973) led to the recognition of the pronounced gravity low associated with the low-density Mesozoic rocks in the Bristol Channel Syncline (Figures 4, 7). Re-interpretation of the Exmoor Gradient Zone in the light of these new gravity data produced some modification to the thrust model (Brooks and Thompson, 1973). The discovery of a high-velocity refractor, identified as the top of the Lower Palaeozoic/ Precambrian, at a depth of only about 2–3 km beneath the north Devon coast (Brooks et al., 1977) resulted in

an alternative model in which the Exmoor Gradient Zone was explained, without invoking thrusting, by assuming that Lower Palaeozoic/late Precambrian rocks had a low density and formed the core of the west-north-west-trending Lynton Anticline (Figure 8).

Superimposed on the Exmoor Gradient Zone are a series of shorter wavelength variations which can be related to density differences between the arenaceous and argillaceous members in the Devonian sequence (compare with Edmonds and Williams, 1985; Cornwell, 1986). These variations are recognisable on the residual gravity map (Figure 6) as a series of elongated features parallel to the strike. Positive anomalies occur over the denser slate formations and negative anomalies over the sandstones.

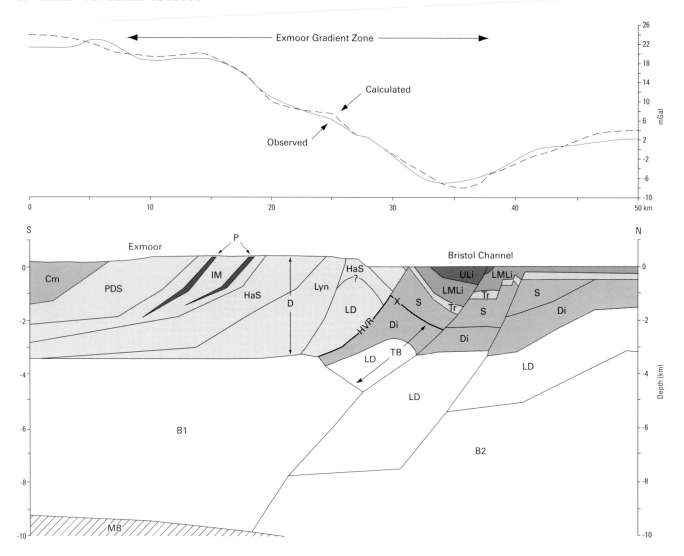

**Figure 7**   Bouguer gravity anomaly profile AA′ (see Figure 4 for location) and model producing calculated profile shown.

Density ranges (in Mgm⁻³) for components of model: Mesozoic rocks (2.45–2.55): ULi–Upper Lias and younger rocks, LMLi–Lower and Middle Lias, Tr–Triassic; Carboniferous: Di–Dinantian (2.70), S–Silesian (2.50), Cm–Culm (2.65); D–Devonian (north Devon)(2.55–2.70), (Lyn–Lynton Formation, HaS–Hangman Sandstone, IM–Ilfracombe and Morte Slates, PDS Pickwell Down Sandstones), P–magnetic beds in Devonian (2.70), LD–Lower Devonian/pre-Devonian low density rocks (2.60); B1–basement 1 (2.78), B2–basement 2 (2.70), MB–magnetic basement (2.78 and 2.73).
X–seismic reflector, HVR–high velocity refractor, TB–thrust block (shallow high velocity layer).
Background field 0 mGal, background and Bouguer correction density 2.70 Mgm⁻³.

The main feature relevant to the district is G1 (Figure 8) which is related to the presence of the Lynton Formation in the core of the Lynton Anticline. South of the district, feature G2, around the Hangman Sandstone–Ilfracombe Slates boundary, is displaced dextrally by the Timberscombe Fault.

Aeromagnetic anomalies provide information on the nature of the pre-Devonian 'basement' of the district. The long wavelength component of the magnetic field shows a gradual increase in values eastwards across the district (Figure 5); these correspond to changes in the depth to a deep-seated magnetic basement of uncertain origin. This eastward increase is almost normal to the axis of the Mesozoic basin in the Bristol Channel and is also apparently undisturbed along the trend of the Exmoor Gradient Zone, implying that the magnetic source lies below those features. Mechie and Brooks (1984) noted that, as the rise of the seismically defined crystalline basement from 7.5 km to 2.3 km is not reflected by the aeromagnetic data, it is likely that the upper part of this basement is non-magnetic. This unit could comprise metamorphic rocks of the basement (Brooks et al., 1993), but the view taken here is that, in places, it may be Carboniferous Limestone.

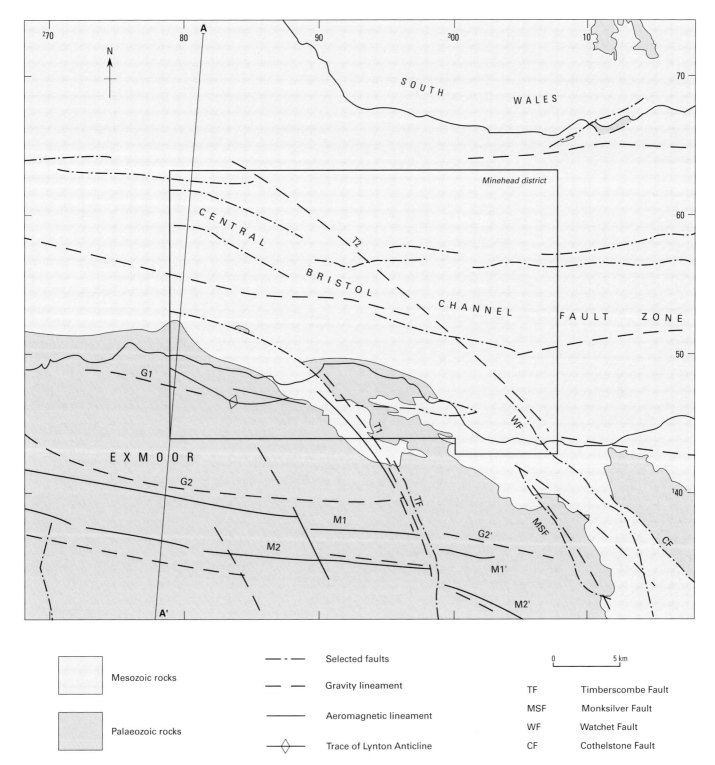

**Figure 8**   Compilation map showing main geophysical lineaments of the district and adjacent areas.

G–lineaments defined on gravity data.
M–lineaments defined on magnetic data.
T1, T2–south-east-trending lineament (see text, p.24, for explanation).

Two main magnetic horizons cross the crop of the Devonian rocks south of the district and are marked by parallel linear magnetic anomalies about 3 km apart (Figure 5). The more prominent, northern, anomaly occurs near the base of the Ilfracombe Slates; the second anomaly lies near the top of the Morte Slates. These anomalies do not everywhere parallel exactly the geological strike and are greatly reduced in size approaching the Timberscombe Fault. East of that fault, the magnetic zone can be traced only by low amplitude anomalies slightly offset to the south-east (M1' and M2' in Figure 8). If the source of the anomalies is stratigraphically related, they might be expected to follow the nose of the Lynton Anticline (Figure 8) and, to the east of the Timberscombe Fault, trend north-westwards towards the Minehead district. However, there is no evidence for this and it seems likely that the formation and distribution of the pyrrhotite which gives rise to the anomalies was controlled by late- or post-Variscan faulting.

## Porlock Basin

Post-Variscan basins, because of the significant density contrast between their infill and the surrounding rocks, are commonly associated with negative gravity anomalies. The basin, occupied by Permo-Triassic and Jurassic rocks, that extends inland from Porlock Bay would, therefore, be expected to produce a Bouguer gravity low, and on the regional gravity map this appears to be indicated by deflections in the contours (Figure 4). The regional gravity coverage of the Vale of Porlock was increased by establishing 24 infill stations and two detailed traverses, totalling 4.4 km in length, between Bossington and Luckbarrow [SS 8985 4810–8920 4613] (traverse 1–1' of Figure 9) and between Dean's Cross and Holt [SS 9264 4688–9235 4478] (traverse 2–2' of Figure 9). Using data from both the regional survey and the additional infill and detailed surveys, residual anomaly maps (Figures 6, 9) have been prepared by removing regional fields. They show an elongated gravity low extending along the basin, with local minima near the coast, where the amplitude is about -1.5 mGal, and in the extreme south-east. The residual anomaly maps, however, also show comparable variations of amplitude within the surrounding Devonian rocks, notably the paired high and low to the west, reflecting the contrast noted earlier between arenaceous and argillaceous lithologies.

On the basis of gravity evidence (Figure 9), the sedimentary infill of the Porlock Basin is probably thickest at the coast. The amplitude of the gravity low decreases south-eastwards into the area around [SS 925 450]. The gravity evidence indicates a ridge across the narrow part of the basin at about [SS 925 440]. In the extreme south a local gravity low centred at [SS 940 425], near Wootton Courtenay, is probably due, in part at least, to the effect of low density Devonian sandstones. The form of the anomalies on the two detailed gravity profiles suggests that the basin is asymmetrical, with a steeper north-eastern margin. Quantitative interpretation is complicated by the presence of anomalies associated with the Devonian rocks and by the limited range of rock densities, but a simpli-

fied model, based on the Bossington–Luckbarrow traverse, is presented in Figure 9b. The low amplitudes of the anomalies associated with the Permo-Triassic reflect the relatively high densities of these rocks (Table 4); the Luccombe Breccia, for example, would be difficult to distinguish, using gravity data, from arenaceous Devonian rocks. No densities are available for the Mercia Mudstone in the Porlock Basin but these could also be high (Table 3).

The gravity evidence for the form of the Porlock Basin remains ambiguous and it seems probable that some, at least, of the observed anomalies are associated with density contrasts within the underlying Devonian bedrock.

The Porlock Basin appears to be associated with low amplitude (10–15 nT) aeromagnetic anomalies (Figure 5), the origin of which is not clear.

## Minehead Basin

The distribution of Permo-Triassic rocks in the Minehead Basin coincides with a weak elongate low on the regional gravity map (Figure 4) and there is some indication, from the slight increase in the gravity values towards the coast, that the northern margin of the basin extends offshore towards the eastern edge of the district. The presence of a gravity gradient just west of the western end of the Permo-Triassic crop at Minehead (Figure 4) suggests the presence of a significant east–west structure affecting the Devonian rocks, in addition to controlling the minor occurrences of younger rocks west of Minehead.

## SUB-MESOZOIC GEOLOGY AND STRUCTURE

The presence of Precambrian basement rocks beneath much of the Bristol Channel and South Wales has been inferred from seismic refraction surveys which indicate the presence of a basal refractor with a velocity of about 6.3 km s$^{-1}$ (e.g. Bayerley and Brooks, 1980; Mechie and Brooks, 1984). This refractor, L4 in Figure 10, lies at depths of 5.5 to 6.5 km beneath the Bristol Channel Basin, and at a predicted depth of about 7.5 km beneath the north Devon coast.

Deep reflectors on the offshore seismic reflection profiles are difficult to interpret. Seismic reflection data indicate the presence of a top Lower Cambrian reflector, with characteristic banded reflectors below, in the offshore part of the Minehead district, for example, on profile F (Figure 11) where it is at depths of more than 5 km. This

**Figure 9    a.** Residual Bouguer gravity anomaly map of the Porlock Basin area with contours at 0.25 mGal intervals, after removal of a regional field defined by a fifth order polynomial. Locations of detailed gravity traverses (1–1', 2–2') and extent of Permo-Triassic and Jurassic rocks indicated. See Figure 6 for location of the area in relation to the district.
**b.** Residual gravity profile on traverse 1–1', based on detailed traverse and regional data (data reduction values: Devonian 2.65 Mgm$^{-3}$, Mesozoic 2.45 Mgm$^{-3}$).

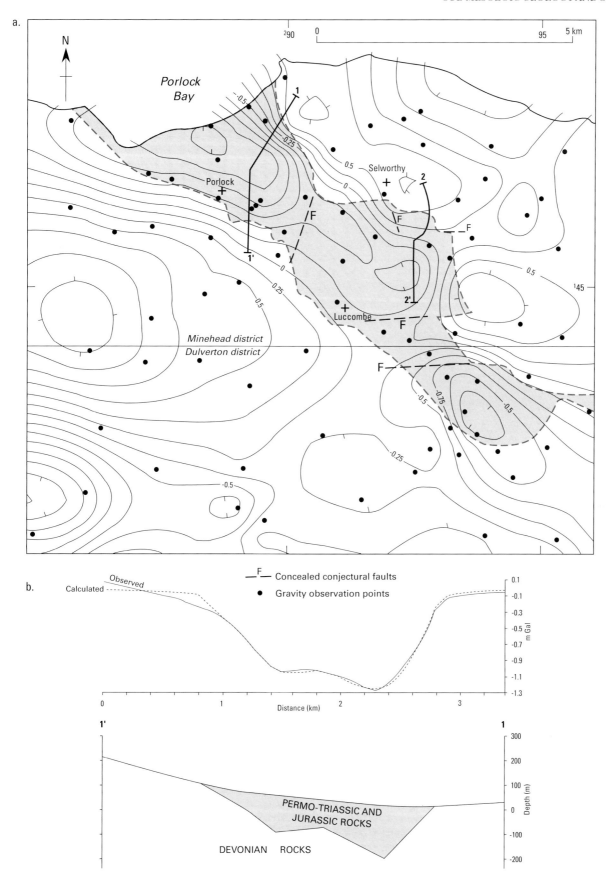

a.

N

Porlock
Bay

²90        0                                  95    5 km

Porlock

Selworthy

Minehead district
Dulverton district

Luccombe

¹45

—F—  Concealed conjectural faults

●  Gravity observation points

b.

Observed
Calculated

0.1
-0.1
-0.3
-0.5   m Gal
-0.7
-0.9
-1.1
-1.3

0                1                2                3
Distance (km)

1'                                                    1

300
200
100   Depth (m)
0
-100
-200

PERMO-TRIASSIC AND
JURASSIC ROCKS

DEVONIAN   ROCKS

a.

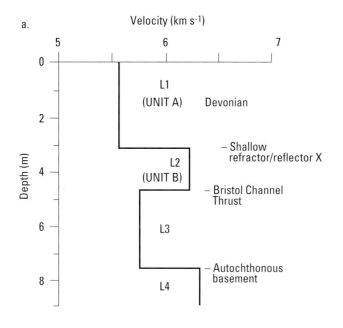

b.

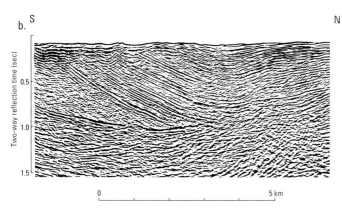

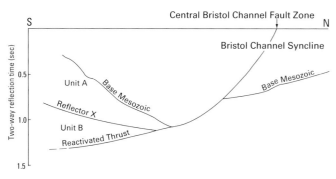

**Figure 10   a.** Velocity-depth model for west Somerset, north Devon and the southern Bristol Channel, after Mechie and Brooks (1984), and interpretation after Brooks et al. (1993); L1–L4 correspond to seismically defined layers recognised by Mechie and Brooks.
**b.** Seismic profile A with geological interpretation, after Brooks et al. (1993); the location of the profile is shown on Figure 11. Seismic data provided courtesy of Geco-Prakla.

reflector has similar characteristics to the top Lower Cambrian reflector seen on a seismic reflection profile which extended from south of Cirencester to the Vale of Pewsey (Chadwick et al., 1983), and which was correlated with Early Cambrian quartzites similar to those at outcrop in the Malvern Hills; the reflector rises west of the district. Ordovician rocks, particularly Tremadoc strata, are likely to be present in the offshore part of the district, judging from the subcrop to the east (British Geological Survey, 1985) and their presence in a borehole in Carmarthen Bay to the west (Tappin and Downie, 1978). Some of the later Ordovician rocks are likely to have been removed before deposition of the Upper Llandovery and later Silurian sediments (Cope, 1987) but they are present in south Dyfed to the west. Silurian rocks, cropping out in the core of a pericline in the Tortworth Inlier, in the Mendips (Tunbridge, 1986), at Cardiff (Waters and Lawrence, 1987), and interpreted on seismic reflection profiles in South Glamorgan (Brooks et al., 1994), have not been recognised in the Minehead district. They may have been removed, providing the source for Lower and Middle Devonian conglomerates in South Wales and Devon (Tunbridge, 1986). Lower and Middle Devonian rocks are also probably missing from the Bristol Channel (Tunbridge, 1986; Bassett and Cope, 1993).

East–west onshore and nearshore seismic refraction lines show a high-velocity refractor at much shallower depths than the basal refractor (L4) referred to above, and interpreted by Mechie and Brooks (1984) as Precambrian basement. The shallower refractor (L2, Figure 10a) is at a depth of about 4 km at Minehead and rises westwards to about 1 km near Lundy Island. Mechie and Brooks (1984) interpreted these results as evidence for a velocity inversion, with the shallow high-velocity layer (L2) overlying a lower velocity (and lower density) layer (L3, Figure 10a). This in turn was considered to rest on the deep autochthonous basement (L4) detected from the north–south refraction lines.

The results of later offshore seismic reflection surveys provided evidence for the existence of a major Variscan thrust in the Palaeozoic basement beneath the Mesozoic Bristol Channel Syncline (Brooks et al., 1988). Reactivation of this thrust was considered to have been responsible for the formation of the Central Bristol Channel Fault Zone. The thrust is not well imaged on seismic profiles in the Minehead district, but is well seen on profiles to the west (Brooks et al., 1988, fig. 3) where it forms a reflection event dipping southward to two-way travel times of more than 2.0 s.

Subsequently, a prominent reflector was recognised on some profiles in the hanging-wall block of the Bristol Channel Thrust at two-way travel times of about 0.7–1.1 s (Brooks et al., 1993). This reflector ('Reflector X') is well developed on a profile (B) along the western boundary of the district (Figures 11, 12a, b) where it separates an upper, rather featureless, unit from one characterised by irregular, laterally discontinuous reflection events. Brooks et al. (1993) considered that Reflector X and the shallow

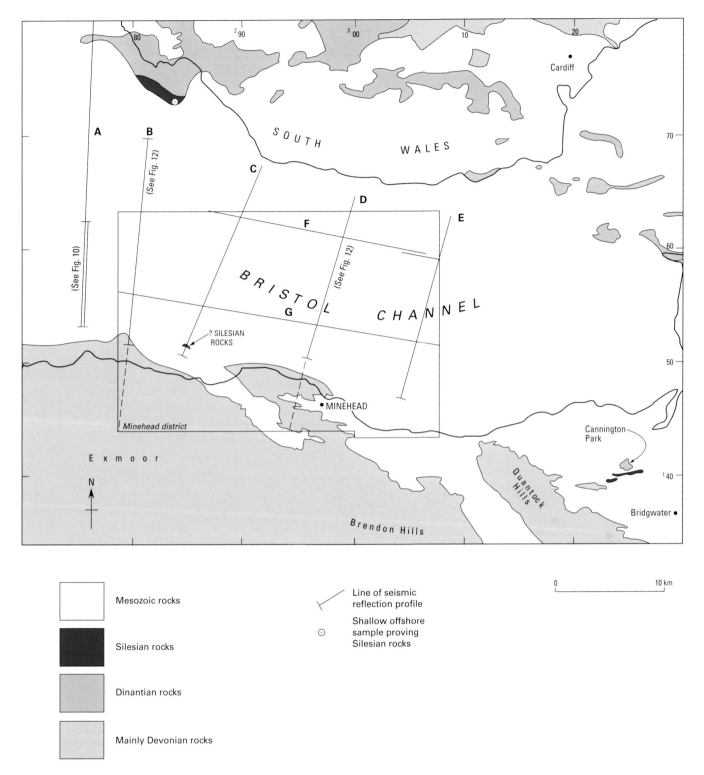

**Figure 11**   Map showing location of seismic reflection profiles relevant to the Minehead district and generalised geology of the district and adjacent areas. Seismic data provided courtesy of Geco-Prakla.

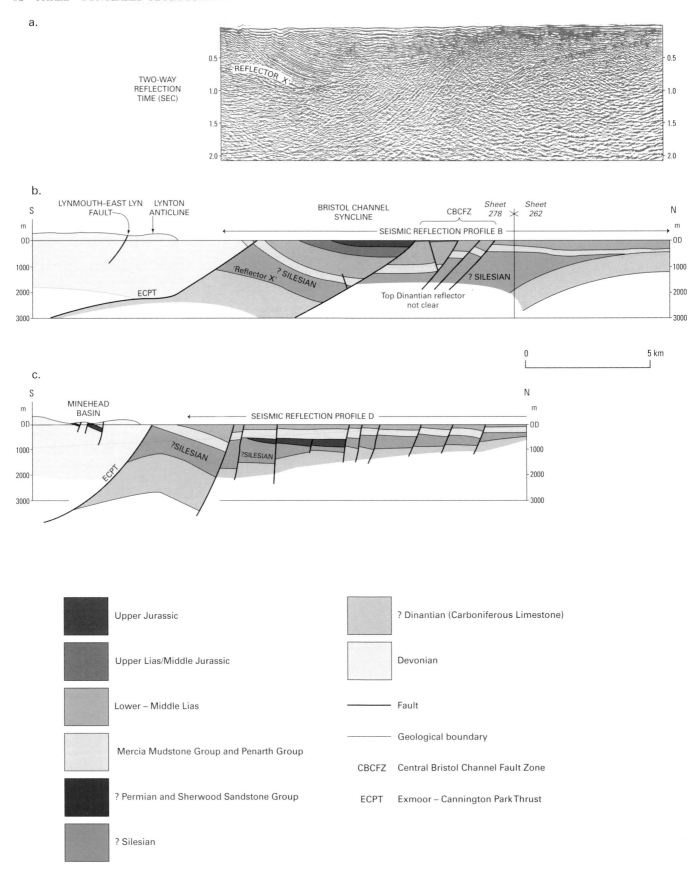

**Figure 12** **a.** Seismic reflection profile B.
**b, c.** geological interpretation of profiles B and D, incorporating contiguous onshore areas.
Location of profiles indicated on Figure 11. Seismic data provided courtesy of Geco-Prakla.

high-velocity refractor present at a depth of about 4 km at Minehead represent a single interface extending beneath north Devon and west Somerset and along the southern Bristol Channel; it is truncated to the north against the south-dipping Central Bristol Channel Fault Zone (Figure 12). Its present geometry is affected by deformation associated with the formation of the Mesozoic Bristol Channel Basin. Corrections for the effects of this deformation (Brooks et al., 1993, fig. 5b) indicate that the surface originally dipped south at 15–20° under the southern part of the Bristol Channel.

The velocity-depth model suggested by Brooks et al. (1993) is summarised in Figure 10a, together with the relationship of the reflection events, as interpreted from seismic profile A, just west of the district (Figure 11). The interface represented by Reflector X and the shallow high-velocity refractor was considered by Brooks et al. (1993) to be a single geological boundary of major significance, but, in the absence of boreholes, the lithology and age of the layers above (L1) and below (L2) this surface are uncertain. Brooks et al. (1993) considered that there were no major lithological units younger than the outcropping Devonian rocks that could be reasonably correlated with the underlying high-velocity layer (L2/Unit B, Figure 10). They proposed that Unit B represented a major lithostratigraphical division older than the outcropping Devonian succession and in normal stratigraphical contact with it, and suggested that it could be sub-Devonian basement schists or gneisses.

An alternative explanation for Unit B, adopted in this memoir, is that it represents the Carboniferous Limestone. Among the evidence favouring this view are sonic log data which indicate that the velocities of 6.2 km s$^{-1}$ in Unit B can be matched in the Carboniferous Limestone elsewhere in South Wales and southern England. For example, in the Knap Farm Borehole [ST 2479 4011] at Cannington Park, Somerset, Arundian to Tournaisian rocks have an interval velocity of 6.0 km s$^{-1}$ from 150 to 800 m below surface, with the interval between 630 and 700 m having velocities of 6.42 km s$^{-1}$ (Edmonds and Williams, 1985). In the Maesteg Borehole [SS 8528 9245] in South Wales, the Carboniferous Limestone at 800–1300 m below surface has an average interval velocity of 5.78 km s$^{-1}$, with the middle 400 m of the succession averaging 6.0 km s$^{-1}$. A number of other boreholes in the Wessex Basin, where the Carboniferous Limestone is currently buried between depths of 300–1900 m, have velocities greater than 6.1 km s$^{-1}$ over an interval amounting to more than half the Dinantian rocks penetrated. This evidence for high Carboniferous Limestone velocities is contrary to the suggestion of Brooks et al. (1993) that there are no lithological units younger than surface Devonian rocks with velocities high enough to be correlated with Unit B.

Interpretation of Unit B as Carboniferous Limestone is consistent with the location of the Bristol Channel along-strike from the known onshore outcrops of Carboniferous Limestone at Cannington Park and in the Mendips, and with the known presence of Silesian rocks in the Bristol Channel. Palaeogeographical considerations make it likely that Dinantian rocks are present along-strike beneath the Mesozoic rocks of the Bristol Channel (Whit-

taker, 1978a). In contrast, the hypothesis of Brooks et al. (1993) implies that Carboniferous rocks are missing from large parts of the Bristol Channel.

Further circumstantial evidence for the presence of Carboniferous Limestone beneath the district is provided by the character and distribution of seismic reflectors. A package of shallow sub-Mesozoic reflectors at the northern end of line B (Figure 12 a,b) can be tied to onshore outcrops of Carboniferous Limestone in South Wales. A well-defined reflector at the top of the package was identified as the top of the Carboniferous Limestone by Brooks et al. (1988, fig.4); it extends to approximately the northern boundary of the district where it ceases to be resolved. This reflector equates with reflector L in South Glamorgan (Brooks et al., 1994). A parallel, but lower and less well-defined reflector, is possibly the base of the Carboniferous Limestone, indicating a thickness for that unit of about 850 m.

The strata overlying the postulated Carboniferous Limestone are presumed to be Silesian in age. Silesian, probably Westphalian, rocks were proved offshore near Porthcawl, north of the district (Figure 11). There is difficulty in tracing the possible top Carboniferous Limestone reflector southwards across the large faults of the Central Bristol Channel Fault Zone. However on the south side of this zone, a reflector with similar character is present and is there identified with Reflector X of Brooks et al. (1993). Reflector X may, therefore, mark the top of the Carboniferous Limestone and be overlain conformably by Silesian strata (Figure 12). Seismic profile C indicates that possible Silesian strata, lying above Reflector X, have a small outcrop [SS 847 514] at the sea bed about 3 km offshore and 6 km north-west of Porlock (Figure 11). The interpreted Variscan thrust which passes up into the Central Bristol Channel Fault Zone is not clearly resolved on profile B, but on a profile (A) west of the district the thrust lies below Reflector X (Figure 10b).

Beneath the onshore part of the district, the shallow refractor lies at depths of 2–4 km (Brooks et al., 1993, fig. 5) and dips southward. Reflector X, offshore, is considered to represent the top of the Carboniferous Limestone, as discussed above; the onshore refractor, which is correlated with Reflector X, is thought to represent the thrust truncation of the Carboniferous Limestone (Figure 12). Such a thrust may be the along-strike westward extension of the Exmoor–Cannington Park Thrust. The putative thrust is thought to have undergone Mesozoic reactivation and to have acted as a normal fault in post-Variscan sequences, at the southern ends of seismic profiles B and D (Figures 11, 12).

## POST-VARISCAN GEOLOGY AND STRUCTURE

Permian to early Cretaceous crustal extension and lithospheric thinning resulted in the formation of thick sequences of sedimentary rocks in fault-bounded basins across north-west Europe. Subsidence of the basins was interrupted by local periods of uplift. In the district, the east–west-trending Bristol Channel Syncline is a major Mesozoic structure containing about 2000 m of strata up to late Jurassic in age and extending for about 200 km from south of Dyfed in the west to Somerset in the east (Kamerling, 1979). Within the Bristol Channel, the Mesozoic rocks occupying the syncline are flanked to the north and south by upper Palaeozoic rocks folded during the Variscan Orogeny. The Bristol Channel Syncline flattens out in the centre of the offshore area; east–west profiles indicate that there is a saddle between it and the Central Somerset Basin towards the eastern side of the district; the crest of the saddle is most clearly seen on line G (Figure 11), at about [ST 040 520].

Between profiles C and D (Figure 11), deeper uncon-formable reflectors may be related to Permian and Sher-wood Sandstone Group rocks, which are not present farther west, but occur to the east in the Burton Row Borehole [ST 3356 5208] (Whittaker and Green, 1983).

The seismic reflection profiles indicate that, in the present district, the Bristol Channel Syncline is an asym-metrical structure with a southern limb dipping gently to the north, and a more complex, faulted northern limb (Brooks et al., 1988); the structure is well displayed on north–south seismic profile B, close to the western edge of the district (Figures 11, 12a, b), where it contains about 2 km of Mesozoic (Triassic to Upper Jurassic) strata. The faults cutting the core and northern limb of the syncline hade to the south and merge at depth into a southerly dipping interpreted thrust (the 'Bristol Channel Thrust' of Brooks et al., 1993), traceable to a depth of at least 6 km. These faults collectively form part of the Central Bristol Channel Fault Zone (Figure 12) and are inter-preted as being caused by reactivation of a Variscan thrust during late Jurassic/early Cretaceous times (Brooks et al., 1988). A detailed interpretation of profiles farther west, between Carmarthen Bay and Lundy Island, shows a reg-ular pattern of north-north-west-trending faults cutting the east–west faults of the Central Bristol Channel Fault Zone. The north-north-west faults probably have a com-ponent of Cainozoic movement, but may have originally acted as transfer faults during the Mesozoic, breaking the north–south extension along the Central Bristol Channel Fault Zone into a number of discrete blocks.

Brooks (1987) suggested that deposition of Triassic to Upper Jurassic rocks in the Bristol Channel Syncline was influenced by crustal extension producing movement on the Central Bristol Channel Fault Zone, which acted as a listric normal growth fault. In this model, the southern limb of the syncline was interpreted as part of a rollover structure, with the implication that the syncline was a basin which reflected the results of syndepositional move-ment on the listric fault. However, Brooks et al. (1988) noted the parallel nature of the outcrop of the Mesozoic strata and the absence of significant onlap and local thickening of the units, indications that the area of the present syncline was part of a much larger basin which covered much of South Wales and south-west England. They inferred that the present Bristol Channel Basin is a post-Jurassic feature and that the folding and fault move-ment which led to the development of the structure occurred in the early Cretaceous.

Beach (1987) suggested that many of the early Mesozoic extensional fault zones in north-west Europe underwent compression and reversal of movement during Cainozoic inversion; an example being '...the compressional and reverse faulted zones along the south side of the Bristol Channel Basin uplifting Devonian against Jurassic' (Beach, 1987, p.47).

Structural studies of the outcrops of Late Triassic and Early Jurassic rocks at the southern and northern mar-gins of the Bristol Channel Basin (Nemčok et al., 1995; Dart et al., 1995) revealed the following three phases in the tectonic evolution of the area:

1. *Permian–Lower Cretaceous extension.* North–south-oriented stretching produced a well-developed system of extensional faults with mainly east–west trends. Many faults probably resulted from reactivation of Variscan thrusts.

2. *Lower Cretaceous–Cainozoic inversion.* North–south-oriented contraction resulted in partial inversion of the fault systems formed during phase 1, with the development of hanging-wall buttress anticlines and zones of intense folding.

3. *Cainozoic strike-slip faulting.* Continued north–south compression resulted in north-west-trending dextral and north-east-trending sinistral strike-slip faults. The most prominent of these structures in the district is the north-west-trending Watchet Fault (Figure 8).

# FOUR

# Devonian

The oldest rocks which crop out in the district are of Devonian age; two lithostratigraphical divisions are present (Figure 13). The Lynton Formation (Lynton Slates of Edmonds et al., 1985) is the oldest Devonian formation exposed in north Devon and west Somerset and ranges in age from late Emsian to earliest Eifelian (Evans, 1983). It is overlain by the Hangman Sandstone Formation (Hangman Grits of Edmonds et al., 1985), which ranges in age from Eifelian to possibly early Givetian.

No samples of Devonian strata have been recovered in the offshore area, because the sea-bed sediments are relatively thick near the coast and were not penetrated by the gravity corer during sampling, and because recovery of samples by this method is unlikely, owing to the hardness of the sandstones. Thus, there are uncertainties in the placing of the boundary between the Devonian and Mesozoic rocks, which is positioned on the map landward of the limits of seismic profiles and the location of Mesozoic samples. The boundary may be closer to shore than shown, but there is presently no means of determining its position more precisely.

## LYNTON FORMATION

In this memoir, the term Lynton Formation replaces the informal name 'Lynton Slates' used in the adjacent Ilfracombe (Sheet 277) district (Edmonds et al., 1985). Earlier names for this unit include 'Lynton Beds', 'Linton Calcareous Group' and 'Linton Group' (Simpson, 1959). The

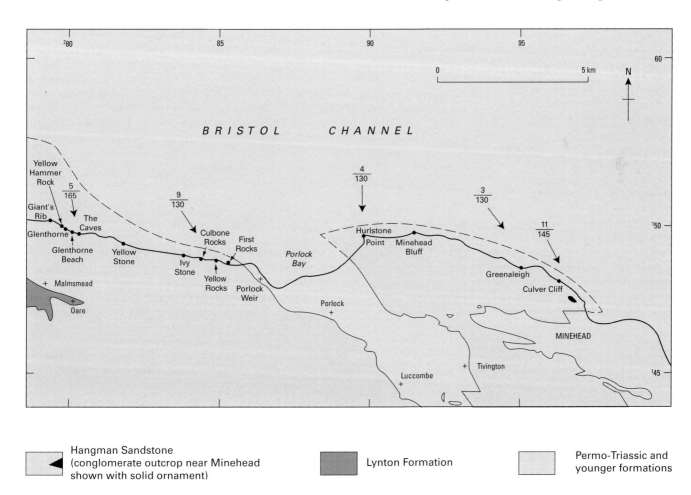

**Figure 13**  Distribution of Devonian rocks in the district, showing the main localities referred to in the text.

The approximate offshore limits of the Hangman Sandstone are shown by broken lines. Faults are omitted. Palaeocurrent vectors, shown by arrows, are after Tunbridge (1984, fig. 2). The upper figure refers to the number of measurements made at a locality; the lower figure refers to the vector mean.

name 'Lynton Slates' is considered unsuitable because the term is not formally defined, and includes a lithological description that is inappropriate since slate forms a very small part of the succession, which is dominated by finely laminated sandstones and mudstones (Evans, 1983).

The Lynton Formation has only a small outcrop area (less than 1 km²) in the district, between Oare and Malmsmead, and along the valley of Badgworthy Water (Figure 13). About the uppermost 250 m of the formation is estimated to crop out within the district and the few exposures indicate that this part consists mainly of grey slates and slaty sandstones, exposed intermittently along the Badgworthy Water between Malmsmead and Cloud Farm [SS 794 467]. A small quarry [SS 7931 4746] about 300 m south-south-east of Malmsmead Bridge shows up to 8 m of the formation, comprising 4 to 6 m of grey slate overlying 0.9 m of grey, cleaved, fine-grained sandstone, on 1 m of grey, slaty sandstone and sandy slate, with pods of fine-grained sandstone. The conformable contact between the Lynton Formation and the Hangman Sandstone is exposed [SS 7935 4672] in the Badgworthy Water west of Cloud Farm. The topmost Lynton Formation consists of blue-grey slates with slaty sandstones near the contact with the Hangman Sandstone, which consists of convolute-bedded sandstones with intercalated slates.

The formation occurs in the faulted core of the Lynton Anticline (p.93), and is overlain conformably to the south by Hangman Sandstone dipping south, mainly between 20 and 35°. To the north, it is faulted against the Hangman Sandstone along the Lynmouth–East Lyn Fault (p.93). The base of the Lynton Formation is not exposed in the district.

Simpson (1964) described the succession to the west, in the adjacent Ilfracombe district, as comprising about 1300 feet (400 m) of mainly finely laminated, fine-grained sandstones in beds varying from 0.3 or 0.6 m up to 3 or 4.5 m thick, with occasional beds of 'pure' slate or sandstone and several thick packets of 'fairly pure' slate in the highest part of the succession. Evans (1983) noted that the sequence is dominated by finely laminated sandstones and mudstones. Edmonds et al. (1985) divided the 'Lynton Slates' into a mainly arenaceous lower part, 200 m thick or more, and a higher part, 120 to 150 m thick, of slates up to 30 m thick with intercalated sandstones similar to those in the lower division. At the top, beds of cleaner sandstones herald the transition into the overlying Hangman Sandstone.

No fossils have been recorded from the Minehead district, but, to the west, in the Ilfracombe district, the trace fossil *Chondrites* is a characteristic feature of the Lynton Formation. Body fossils are mainly concentrated in discrete shell bands; they are mostly distorted or otherwise poorly preserved, and the fauna is of low diversity. Bivalves, brachiopods, bryozoans and crinoid debris predominate, and tentaculitids are locally abundant. Edmonds et al. (1985) noted that, of the bivalves, forms of *Palaeoneilo* and *Ptychopteria* (*Actinopteria*) are relatively common, and that, among the brachiopods, spiriferaceans are common and strophomenoids and orthoids are scarcer. Evans (1983) noted that the brachiopod fauna is remarkably

consistent throughout the succession, and is dominated by *Platyorthis longisulcata*, *Chonetes sarcinulatus*, and *Subcuspidella lateincisa*. The brachiopods indicate a late Emsian age for the succession as a whole, though the uppermost part may be early Eifelian.

Knight (1990) recovered low-diversity, low-abundance, shallow-water (icriodid-dominated) conodont faunas from the Lynton Formation. The micropalaeontology, principally the palynology, of the formation indicates a late Emsian to early Eifelian age.

## Depositional environments

From a study of the more extensive sections of the Lynton Formation to the west, in the Ilfracombe district, Edmonds et al. (1985) deduced that the formation accumulated in a shallow sea with periods of active deposition of sand alternating with quieter periods of mud deposition. This is in agreement with the work of Simpson (1964), who also suggested a similar depositional setting, with the sea bottom affected by wave action only during storms.

No detailed sedimentological study of the Lynton Formation of this district has been carried out. However, it is likely that deposition was marked by fully marine conditions, possibly in an environment that varied from lower shoreface to offshore. The trace fossil *Chondrites* is indicative of fully marine conditions, but provides little indication as to the water depth. Deposition of sand occurred during periods in which higher energy currents could transport arenaceous sediment, with muddy laminae and beds deposited from suspension. Discrete shell beds may represent storm deposits with the coarser bioclastic debris transported offshore. The presence of sandstone beds with wave rippled tops, described by Edmonds et al. (1985), indicates that the sea bottom was influenced at times by oscillatory currents and indicates shallowing to a position above storm wave base.

## HANGMAN SANDSTONE FORMATION

The marine facies of the Lynton Formation was succeeded, in mid-Eifelian times, by a wedge of clastic, continental, alluvial plain deposits which prograded southwards from a source in South Wales (Bluck et al., 1992) and comprises the Hangman Sandstone Formation. The formation consists mainly (about 75 per cent) of purple, grey and green, fine- to medium-grained sandstones, about 36 per cent of which occur as single-storey and multistorey channel sandstones up to 4 m thick, and about 39 per cent are thick and thin sheet sandstones in units mainly less than 1 m thick. The remainder of the sequence is made up of reddish brown mudstones. Eight facies have been recognised in coastal exposures in the district (Jones, 1995) and are described below. The sequence is broadly homogenous; lithological variations are mostly minor and are not indicated on the map, apart from a small occurrence of conglomerate and pebbly sandstone, of uncertain stratigraphical position, mapped around [SS 968 471], about 1 km north of Minehead (Figure 13).

The Hangman Sandstone outcrop, of 72 km², covers about three-quarters of the land area of the district (Figure 13). Between the western boundary of the district, near Glenthorne, and Porlock Weir, there are generally low cliffs or small exposures at the base of the heavily wooded coastal slope, except where landslip has obscured exposures or where head is present, as at Embelle Wood. The best exposures along this part of the coast are around Glenthorne, which is a Site of Special Scientific Interest (SSSI) for the Hangman Sandstone (p.19), and at Yellow Stone, Ivy Stone, Culbone Rocks, Yellow Rocks and First Rocks. Between Hurlstone Point and Minehead, the coast is mainly unwooded, but extensively landslipped. The best exposures of the Hangman Sandstone along this section of the coast are the cliffs around Hurlstone Point, around Minehead Bluff, at Greenaleigh and at Culver Cliff. Selected coastal exposures typifying the formation are described and illustrated below (pp.48–51).

Inland, the Hangman Sandstone forms hills which rise to over 400 m above OD and comprise a mixture of heather moorland, farmland and woodland forming part of eastern Exmoor. Here, exposures occur mainly along the larger streams; the intervening areas are largely mantled by blanket head and regolith (Chapter 8) and there are few exposures except in trackside cuttings and in scattered disused quarries.

The sandstones north of the Lynmouth–East Lyn Fault were formerly included in the Foreland Grits and were thought to occur stratigraphically beneath the Lynton Formation (e.g. Champernowne and Ussher, 1879). However, the Foreland Grits are now regarded as part of the Hangman Sandstone (see Edmonds et al., 1985, pp.9–10).

The term 'Hangman Sandstone Group' was used by Tunbridge (1978) for this sequence in north Devon and west Somerset; he divided it into five formations, as follows:

| | *Thickness* m |
|---|---|
| **Hangman Sandstone Group** | |
| Little Hangman Formation | about 100 |
| Sherrycombe Formation | 90 |
| Rawn's Formation | 148 |
| Trentishoe Formation | about 1250 |
| Hollowbrook Formation | 70 |

These subdivisions cannot be mapped in poorly exposed inland areas in the Ilfracombe district (Edmonds et al., 1985), nor in this district, and therefore they do not satisfy the requirements of a formation. Accordingly, the status of the 'Hangman Sandstone Group' is here amended to that of formation, and the 'formations' of Tunbridge (1978) are classed as members of that formation. The term 'Hangman Grits', used in the Ilfracombe district by Edmonds et al. (1985), has been replaced by Hangman Sandstone, firstly, because the formation consists mainly of fine- to medium-grained sandstone, and, secondly, because the term 'grit' is not used in modern classification schemes.

The bulk of the Hangman Sandstone sequence in the district probably belongs to the Trentishoe Member (Trentishoe Formation of Tunbridge, 1978). None of the other members has been definitely recognised, although conglomerates in the Minehead area, and possibly pebbly sandstones in the coastal exposures at and east of Minehead Bluff [SS 915 494], could belong to the Rawn's Member (Rawn's Formation of Tunbridge, 1978).

Facies analysis (see below) indicates deposition on the distal part of an alluvial fan, with periodic development of an ephemeral continental mudflat. Palaeocurrent data based on measurements of cross-bedding in the Trentishoe Member indicate derivation from the north (Tunbridge, 1984, fig. 2; Figure 13). Tunbridge (1986) has shown that the lithic arenites of the Trentishoe Member closely match those in the higher part of the Lower Old Red Sandstone in South Wales. He considered that the member was derived mainly from the highest Lower Old Red Sandstone in early mid-Devonian times as a result of end-Caledonian uplift and erosion in South Wales. By contrast, the coarse-grained, locally conglomeratic, Rawn's Member (not definitely known to crop out in the Minehead district) has a distinctive assemblage of subrounded to angular clasts of porphyry, tuff, quartzite and lithic arenite which do not match rocks from possible source areas in South Wales. Tunbridge (1986) proposed that the Rawn's Member was derived from an intermittently elevated landmass of Precambrian and Lower Palaeozoic rocks in the Bristol Channel area ('Pretannia' of Cope and Bassett, 1987).

The formation was folded on approximately east–west axes during the Variscan Orogeny, and cleavage is locally developed in the more argillaceous units. Fracturing is widely developed and is locally intense.

Estimates of the thickness of the formation vary, owing to the lack of marker bands in a thick, generally rather monotonous, sequence and, therefore, it is difficult to estimate the effects of folding and faulting. Evans (1922) gave a thickness of 1097 m, Lane (1965) estimated 1350 m, and Tunbridge (1978) 1658 m. Tunbridge's estimate includes the Hollowbrook Formation, not distinguished by Evans or Lane. From calculations based upon dip and outcrop width, Edmonds et al. (1985) considered that up to 2500 m could be present in the adjacent Ilfracombe district, to the west. The full thickness is not exposed in the Minehead district but similar calculations indicate that at least 1000 m of the Hangman Sandstone crops out there.

## Sedimentary facies

The sedimentary features and facies of the formation have been described by Tunbridge (1978, 1981, 1984). Coastal exposures in the district have been examined by Jones (1995) who recognised eight sedimentary facies (Table 5); these are described and interpreted below. The discontinuous and structurally disturbed nature of the exposures precluded a detailed assessment of the interrelationship of these facies. The coarser-grained facies are sandstone-dominated and comprise channel sandstones (Facies 1, 2 and 3) and sheet sandstones (Facies 4 and 5); these facies are commonly interbedded and separated by only thin mudstones. The finer-grained facies (6 to 8), though

**Table 5**   Summary of the sedimentary facies in the Hangman Sandstone of the district.

| Facies | Main characters | Interpretation |
|---|---|---|
| 1. Single-storey channel sandstone | Erosively based sandbodies, up to 3.5 m thick | High-energy river channels |
| 2. Multistorey channel sandstone | Erosively based, vertically stacked sandbodies, up to 12 m thick | High-energy river channels |
| 3. Laterally accreted channel sandstone | Erosively based sandbodies, up to 1 m thick | Low-energy minor river channels |
| 4. Thick sheet sandstone | Sharp or erosively based beds of sandstone in laterally persistent sheets, up to 1 m thick | High-energy sheetfloods deposited on subaerial mudflats |
| 5. Thin sheet sandstone | Laminae to thin beds of sandstone up to 0.1 m thick, typically interbedded with Facies 7 | Distal or weak sheetfloods deposited into lakes; later modification by emergence and desiccation |
| 6. Massive to laminated mudstone | Mudstones in beds up to 2 m thick; local carbonate nodules. | Deposition from suspension in perennial lakes |
| 7. Desiccated and remobilised mudstone | Mudstones, with common desiccation cracks; typically interbedded with Facies 5. Abundant bioturbation and local carbonate nodules. | Deposition from suspension in ephemeral lakes. Emergence caused drying of sediment surface. Folded and convoluted laminae caused by water escape. |
| 8. Mudstone with extraformational pebbles | Mudstones with scattered quartz pebbles | Cohesive, subaerial debris flows |

mudstone-dominated, include thin sheet sandstones.

Palaeocurrents in the formation vary from south-east to south-west, with a vector mean to the south-south-east (Figure 14).

FACIES 1: SINGLE-STOREY CHANNEL SANDSTONE

Facies 1 forms approximately 20 per cent of the succession and comprises erosively based, pale reddish grey to greenish grey, medium- to coarse-grained sandstones in units up to 3.5 m thick (Plate 3); these extend beyond the limits of exposures and may persist laterally for hundreds of metres. Rare mudstone laminae and beds are present, and impersistent mudstone pebble conglomerates occur locally at the base of sandstone beds. Each unit is internally cross-bedded, with sets of planar to open trough cross-bedding up to 2.8 m thick. Low-angle cross-bedding is common, with well-laminated ('flaggy'), downcurrent-descending foresets. Some examples have convex-up foresets, and all sets have unidirectional palaeocurrent azimuths. Rare examples of plane bedding occur (Plate 4), with a well-developed parallel-lamination and primary current lineation present on bedding. Grading is generally absent, although the top of the facies is locally characterised by an abrupt decrease in cross-bedding set size. The upper parts of sandbodies commonly contain cross-lamination and, in rare instances, asymmetrical ripple form sets.

A sandstone bed at Culver Cliff [SS 9614 4783] contains a set of large-scale asymmetrical folds that verge to the north-east (Plate 5). The fold wavelength is 2.7 m and the amplitude 1.4 m. The overlying and underlying beds are unaffected by folding, and the crests of some of the folds are eroded; the folds are locally slightly convoluted.

The thick sandbodies of this facies were deposited within river channels. Sand was transported by unidirectional tractional currents, and dune-sized bedforms were dominant. Dunes are typically low-flow regime bedforms. However, the Facies 1 sandbodies contain low-angle cross-bedding, a structure indicative of formation close to the lower/upper flow regime boundary (Røe, 1987; Saunderson and Lockett, 1983). This structure is also a common feature of the deposits of ephemeral channels, with periodic high energy flows. Flow depths were greater than the thickest bedform, and channel depths of up to 3 m are envisaged. Plane bedding formed in the upper flow regime, under occasional high flow and shallow water conditions (Collinson and Thompson, 1989, p.102). Mudstone laminae and beds were deposited from suspension, during periods of low energy. The alternation of high flow velocity bedforms and low flow drapes provides further evidence for ephemeral channel flow. Absence of grading, combined with a rapid decrease in set size at the top of sandbodies, indicates that rapid flow waning and abandonment occurred within these channels.

The large folds at Culver Cliff [SS 9614 4783] (Plate 5) are interpreted as synsedimentary deformation features; a slump origin, down a north-east-dipping palaeoslope, is suggested by the north-east-directed asymmetry of the folds.

FACIES 2: MULTISTOREY CHANNEL SANDSTONE

This facies forms approximately 15 per cent of the succession and is similar to Facies 1, but comprises composite, vertically stacked, channel deposits (Plate 6) up to about 12 m thick and with individual sandbodies up to 4 m thick. The facies is characterised by the presence of

**Figure 14**    Rose diagram of palaeocurrent measurements for the Hangman Sandstone made during the present survey. Number of readings (restored for structural dip) = 22.

multiple erosion surfaces, commonly lined by mudstone pebble lags. The dominant structure is trough cross-bedding, within scour-shaped troughs up to 0.6 m deep; low-angle cross-bedding, cross-lamination and convolute lamination may be present. The top to Facies 2 is usually sharp, although fining-upward sequences occur.

The indications of erosion within and at the base of Facies 2 are evidence of confined flow within channels. The multistorey nature of the facies is the result of channels eroding into earlier deposits; this may reflect changes in the position of one channel with time, or may be the result of the migration of separate channels on a braidplain. The sedimentary structures and lithologies are similar to those in Facies 1 and indicate similar channel processes, pointing to ephemeral flow. Convolute lamination formed by plastic deformation of partially liquified soft sediment (Collinson and Thompson, 1989, p.154). Flow depths were probably similar to those of Facies 1.

FACIES 3: LATERALLY ACCRETED CHANNEL SANDSTONE

This facies forms only about 1 per cent of the succession. It comprises units of fine- to medium-grained sandstones up to 1 m thick, with erosional bases. The dominant feature is the presence of bedding surfaces inclined at low angles (20 to 30°) relative to the top and bottom of the sandstone bed (Plate 7); these bedding surfaces are typically sigmoidal in form, and split the sandbody into inclined lenticular units. Sedimentary structures include cross-lamination and convolute lamination.

Facies 3 is interpreted as the product of deposition in small, shallow channels. The inclined strata are interpreted as lateral accretion bedding, formed by successive deposition on low-angle, side-attached barforms. The channels were probably sinuous and up to 1 m deep. Flow

within the channels was of low energy, capable of generating only ripple-sized bedforms.

FACIES 4: THICK SHEET SANDSTONE

This facies forms approximately 24 per cent of the succession and comprises sheet-like units of dark reddish grey, medium pinkish brown to greenish grey, fine- to medium-grained, (rarely up to coarse-grained) lithic sandstone (Plates 8 and 9), forming beds traceable laterally for tens to hundreds of metres. Individual beds are generally 0.1 to 1 m thick, although thicker units, representing composite stacked sheet sandstone units, are locally present. Individual sheet sandstones in the composite units can be picked out by the presence of mudstone clast-lined bed bases and mudstone laminae. Beds are sharp or erosively based (Plates 8 and 9), and intraformational mudstone clasts occur at the base of some beds; a few mudstone clasts contain sand-filled desiccation cracks. Extraformational quartz pebbles occur in some beds. Small-scale scours, up to 0.5 m wide and 0.2 m deep, are locally cut into the tops of beds (Plate 9); they have draped infills, with thicker laminae present in the axes of the scours.

The main structures in Facies 4 are low-angle, trough and planar cross-bedding; these commonly pass upwards into cross-lamination and climbing ripple cross-lamination. Current and wave-ripple form sets occur on the upper surface of some beds (Plate 10). Low-angle plane bedding with primary current lineation is common, and convolute lamination and oversteepened cross-bedding are developed locally.

The sheet-like, laterally continuous form of the sandstone beds indicates deposition from unconfined flows. Their thickness suggests that the flows were either proximal (and/or axial to source), or large magnitude floods. Where composite units are developed, frequent flooding events are indicated. High energy, turbulent, tractional flows are indicated by the presence of sharp or erosive bases to sandstones, and by the types of sedimentary structures present. Low-angle cross-bedding was formed by high-velocity flows, close to the dune/plane bed transition (Røe, 1987; Saunderson and Lockett, 1983). Plane beds with primary current lineation indicate that, at times, flow occurred under upper flow regime, plane-bed conditions. Rapidly waning currents are suggested by the upward change from plane and cross-bedding to cross-lamination. The scours were formed by late-stage, waning flows eroding into the top of sheetflood deposits. Reworking of the tops of these deposits in small lakes is indicated by the occurrence of wave-ripple form sets. Such lakes probably formed by ponding of water subsequent to sheetflooding, and were ephemeral.

Mudstone clasts lining bed bases were derived by the erosion of intraformational mudstone beds. Desiccation cracks in some of these clasts indicate that, in some instances, the muddy substrate had dried out and that the cracks were filled with sand before erosion took place; such fragmented material would be easy to remove and transport. The occurrence of desiccated muds also indicates that depositional environments varied from lacustrine to subaerial.

## FACIES 5: THIN SHEET SANDSTONE

Facies 5 forms approximately 15 per cent of the succession and is typically interbedded with Facies 7 (desiccated and remobilised mudstone) (see below); it is characterised by laminae to thin beds (up to 0.1 m) of sandstone, some of which are lenticular (Plate 11). The sandstones are mainly very fine grained, up to fine grained, and the beds are sharp-based and commonly fine upwards. They form sheets which generally persist laterally for only a few metres. The dominant structure is unidirectional cross-lamination. Foresets comprise interlaminated sand and silt, and may be inclined at low angles. Current-ripple form sets may occur, with mudstone drapes. Wave-ripple cross-lamination and wave-ripple form sets also occur, but are less common. The wave-ripples are low-amplitude, straight-crested and symmetrical, with rounded crests. Sandstone of this facies typically fills desiccation cracks within Facies 7; these cracks locally cross-cut multiple beds of sandstone and mudstone.

Deposition of Facies 5 was probably from unconfined, tractional sheetfloods. Waning flows are indicated by the fining-upward profiles, with cross-lamination produced by unidirectional currents. Deposition from weak or distal flows is indicated by the thickness of beds. The interlamination of sand and silt on ripple foresets indicates that sedimentation from suspension accompanied bedload transport. The lenticular nature of some of these units is related either to the generation of asymmetrical, current-ripple formation or to reworking into symmetrical ripples by waves. Deposition, as suggested by the presence of wave-ripples, is interpreted to have been in standing bodies of water, probably lakes, the ephemeral nature of which is indicated by the presence of desiccation cracks.

## FACIES 6: MASSIVE TO LAMINATED MUDSTONE

Facies 6 forms approximately 9 per cent of the succession and comprises

**Plate 3** Sedimentary features of the Hangman Sandstone.

A thick sandbody comprising a single-storey channel sandstone (Facies 1), overlying thick sheet sandstones (lower left). The thick sandbody has an undulatory, erosive base; the sheet sandstones have sharp bases and are individually constant in thickness. The rucksack is 0.5 m high. Yellow Hammer Rock [SS 7980 4975] (GS480).

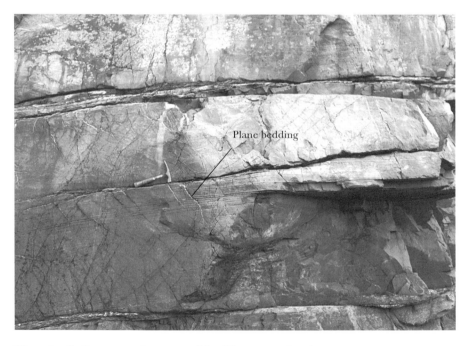

**Plate 4** Sedimentary features of the Hangman Sandstone.

Plane bedding, indicative of upper flow regime currents in Facies 1, immediately beneath the hammer (0.28 m long). Yellow Stone [SS 8195 4911] (GS481).

**Plate 5** Sedimentary features of the Hangman Sandstone.
Synsedimentary folds in Facies 1 reflect slump movement down a north-easterly palaeoslope (to the right). Note the truncation of the crest of one antiform and the undeformed sheet sandstone overlying the folds. The hammer is 0.28 m long. Culver Cliff [SS 9614 4783] (GS485).

**Plate 6** Sedimentary features of the Hangman Sandstone.
Multistorey channel sandstone facies (Facies 2), forming a thick sandbody complex. The wavy/lenticular bedding in the middle unit (Channel 2) reflects trough cross-bedding. Hurlstone Point [SS 8985 4918] (GS486).

beds up to 2 m thick of claystone, silty claystone and siltstone, occurring either separately or finely interlaminated; these lithologies are typically reddish brown or purple, less commonly greenish grey. Reduction spots are also present. The mudstones are massive or laminated, with rare cross-laminated silty lenses; some form the upper parts of fining-upward sequences in which sheet sandstones pass upwards into mudstones. The lack of lamination may be a primary depositional feature; however, at some localities any primary lamination may be obscured by a well-developed cleavage. Bioturbation is locally present in the form of vertical burrows with a clean, finer mudstone infill. Plant stems were noted in a thin (0.02 m) bed of greenish grey, micaceous siltstone in a loose block of Facies 6 at Culver Cliff [SS 9614 4783]. The sediment shows a compactional texture around the stems.

An unusual feature of Facies 6 is the presence of rare dykes filled with grey, structureless, fine-grained sandstone; examples are exposed at Greenaleigh [SS 9521 4816]. They are discordant to bedding, inclined to subvertical in form, typically ptygmatically folded, and irregular in thickness (up to 3.0 m long and 0.03 m wide). At some localities they pass down into undisturbed beds of sandstone.

In Facies 6, the lack of features indicative of subaerial exposure suggests that deposition took place in a perennial body of water, probably a lake. The fine grain size suggests that the sediment was transported in suspension before settling on the lake floor. Low-energy conditions are required for such sedimentation, although the mud may have been supplied in part by high-energy sheet-floods carrying sand and mud together. The formation of lamination was related to subtle fluctuations in sediment supply, the presence of cross-lamination indicating that some of the sediment was transported by tractional processes. The massive appearance of the mudstones may result from i) rapid sedimentation rates which inhibited any sorting of the sediment; ii) deposition as mudflows; or iii) the obliteration of earlier fabrics by bioturbation or cleavage development. Their red coloration is thought to have resulted from an early diagenetic reaction, in which ferrous iron was oxidised to ferric, with the greenish grey coloration representing later reduction.

The sand-filled dykes are interpreted as having formed by the injection of sand from underlying sandstone beds, and in some instances by lateral injection. Liquefaction of the intruding sediment may have been triggered by earthquakes.

### FACIES 7: DESICCATED AND REMOBILISED MUDSTONE

Facies 7 forms approximately 15 per cent of the succession and comprises claystones and siltstones in units up to a few metres thick; it typically occurs interbedded with Facies 5 (Plate 11). Desiccation cracks are commonly associated with small-scale interlamination and interbedding of sandstone and mudstone. Siltstones are dark to medium purple-grey and greenish grey, thinly to thickly bedded, massive or laminated, and fine grained. Claystones are dark reddish brown, thinly to thickly laminated, up to 0.05 m thick, and commonly impersistent laterally.

The dominant feature of Facies 7 is the presence of sandstone-filled desiccation cracks that disrupt the lamination (Plate 12). The cracks locally penetrate multiple beds and laminae, including those of sandstone; in places they are up to 0.3 m long and are straight or ptygmatically folded. The effects of desiccation are locally so intense that the rock is brecciated. Cracking also induced uplifting of claystone beds, causing sand to flow under the upturned edges.

Folded and upturned laminae of claystone are also characteristic of the facies (Tunbridge, 1984, pp.704–706). The laminae locally show small-

**Plate 7**  Sedimentary features of the Hangman Sandstone.

Laterally accreted channel sandstone (Facies 3). Low-angle lateral accretion bedding (arrowed) dips from left to right in the middle sandstone unit. The hammer is 0.28 m long. Ivy Stone [SS 8387 4876] (GS487).

**Plate 8**  Sedimentary features of the Hangman Sandstone.

Thick sheet sandstones (Facies 4). The rucksack is 0.5 m high. Hurlstone Point [SS 8993 4913] (GS488).

scale, chevron-like folds with sharp-crested or tight antiforms and open synforms; in places they clearly drape ripple (possibly symmetrical) form sets; these laminae are locally broken at the crests of the antiforms. Such breakage is commonly associated with water escape and sand-flowage or diapir structures. In certain instances, sand flowage along multiple laminae has caused disruption of entire beds up to 0.5 m thick.

Bioturbation is common within Facies 7, and two types have been recognised. The first consists of burrows that are irregular in width, up to 0.03 m, and many centimetres in length. Some swelling of the walls is present, giving the burrows an undulose form. They are near-vertical to horizontal, with the horizontal component dominant, and are linked by near-vertical to inclined segments with a complex branching, almost polygonal form; a massive, sand- or mud-fill, is common, though heterolithic, retrusive (concave-up) meniscate infills are locally present (Plate 12). Some siltstone clasts occur within the burrow fills. The burrows lack marginal tubes. The second type of burrow is less common, and consists of simple, rounded, horizontal, sand-filled burrows.

The mudstones of Facies 7 were deposited from suspension in a quiet-water, lacustrine environment. The common occurrence of thin sheet sandstones, and mud draping of ripple form sets suggests that most of the mud was also supplied from sheetflood events. The lakes were ephemeral features, subject to repeated drying phases which produced desiccation cracks. The folding of claystone laminae and its association with sand diapir structures has been described by Tunbridge (1984), who proposed that the folding is due to mud behaving in a ductile manner in response to upward-directed forces. These forces are caused by the upward flow of sediment owing to excess pore water pressure in sands (Tunbridge, 1984).

The main type of bioturbation within Facies 7 is thought to be *Bea-*

**Plate 9**   Sedimentary features of the Hangman Sandstone.
Thick sheet sandstones (Facies 4). The lower sheet is sharp-based; two scours (arrowed) cut into the top are filled with a drape which thickens into the scours. The hammer is 0.28 m long. Ivy Stone [SS 8387 4876] (GS489).

**Plate 10**   Sedimentary features of the Hangman Sandstone.
Symmetrical, straight-crested wave ripples on the top of a sheet sandstone (Facies 4). Tuning-fork bifurcations are present in places. The lens cap is 5 cm in diameter. Ivy Stone [SS 8387 4876] (GS490).

*conites*, recent interpretations of which favour an origin either by vertebrate locomotion or as a back-filled arthropod burrow (Eager et al., 1985, pp.137–138). The burrows were formed in a quiet, presumably wet or damp, environment. The second type of burrow is probably *Planolites*, a simple feeding burrow produced by an infaunal worm (Eager et al., 1985).

### FACIES 8: MUDSTONE WITH EXTRAFORMATIONAL PEBBLES

Facies 8 forms less than 1 per cent of the succession and generally consists of beds of massive, pinkish purple, siltstone containing scattered extraformational clasts (Plate 14). The beds are typically from 0.07 to 0.12 m thick, rarely up to a maximum of 1.0 m thick. The extraclasts, of vein quartz and rare quartzite, are typically of medium pebble size but range from granule to large pebble size. They are generally subangular, more rarely angular to subrounded, and are scattered randomly within the massive siltstone. The beds are matrix supported, poorly sorted, and show no obvious grading.

The unusual combination of mudrock with scattered pebbly extraclasts suggests that Facies 8 represents the deposits of subaerial debris flows. The mud-dominated, strongly matrix-supported nature of the beds and lack of grading indicates that the flows were cohesive, with the clasts fully supported by the matrix. The angular nature of the clasts indicates a lack of abrasion and suggests that there was limited movement within the flow.

### POST-DEPOSITIONAL CARBONATE NODULES

Post-depositional carbonates occur in less than 1 per cent of the succession, principally in mudrocks of Facies 6 and 7, but also in heterolithic sandstones and mudstones of Facies 2. They comprise white to pale yellowish brown, subrounded carbonate nodules (Plate 13) elongate parallel to bedding and forming units up to about 1 m thick. The nodules are uncoalesced, and up to 3 × 2 cm in size. They have a displacive texture which is locally

**Plate 11** Sedimentary features of the Hangman Sandstone.

Thin sheet sandstones (white) (Facies 5) interbedded with mudstone laminae (reddish brown) of Facies 7. Note the common occurrence of desiccation cracks within the mudstone, and some upturned laminae. The hammer is 0.28 m long. First Rocks [SS 8520 4850] (GS491).

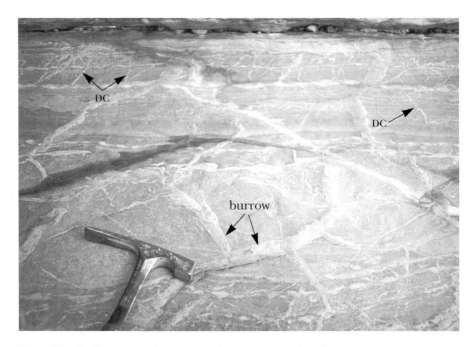

**Plate 12** Sedimentary features of the Hangman Sandstone.

Purple siltstone (Facies 7) with laminae and thin beds of pale grey sandstone (Facies 5) and abundant, thin, sand-filled desiccation cracks (DC). The thicker sand to the right of the hammer has a branching pattern and an internal, retrusive, meniscate structure is visible in places. This represents the burrows of ?*Beaconites*. The hammer head is 0.14 m long. Yellow Hammer Rock, Glenthorne [SS 7980 4975] (GS492).

associated with a concentric layering, so that films of mud are trapped between the nodules. The surrounding mudrock commonly has a disrupted or listric texture. In some instances, the carbonate nodules appear to be nucleated around roots. The nodules are thought to consist of calcium carbonate, and are probably pedogenic in origin, having formed by the precipitation of carbonate, related to high rates of evapo-transpiration of carbonate-rich groundwater (Wright and Tucker, 1991). They represent immature calcrete horizons.

## Petrography

The petrography of sandstones in the Trentishoe Member in and west of the district was summarised by Tunbridge (1986). They are mostly fine- to medium-grained lithic arenites. The bulk of the rock is made up of quartz, consisting mainly of single grains showing undulose extinction, and some polycrystalline quartz. Feldspars, represented by altered orthoclase, microcline, braid perthite and plagioclase ($An_{55-75}$), form about 5 per cent of the rock. Rock fragments are abundant; igneous material includes fragments of dust tuffs, spherulitic grains, acid lava with feldspar microlites, and grains with granophyric intergrowths of quartz and feldspar. Metamorphic rock fragments are less common, and consist of highly sericitized phyllite and chloritic metaquartzite. Sedimentary rock fragments include lithic greywackes and micaceous quartzites.

Thin sections (E 70300–70302, 70304 and 70310) of sandstones from the Hangman Sandstone in the district were examined by Strong (1995). Their detrital mineralogy is dominated by quartz and chert; minor feldspar, mica and relict detrital clay pellets occur. There is abundant secondary opaque material (probably Fe-oxide/hydroxide) and a secondary silica cement.

Conglomerates are rare; small exposures [SS 9657 4724–9680 4713]

**Plate 13**   Sedimentary features of the Hangman Sandstone.

Mudstone with carbonate nodules. The carbonate, representing incipient calcrete development, occurs as white nodules (CN). The lens cap is 5 cm in diameter. First Rocks [SS 8520 4850] (GS493).

**Plate 14**   Sedimentary features of the Hangman Sandstone.

Jointed mudstone with extraformational pebbles (Facies 8). Scattered, angular quartz pebbles are common within a siltstone bed in the lower third of the photograph. The hammer is 0.28 m long. Greenaleigh [SS 9521 4816] (GS494).

are present about 1 km north of Minehead. A thin section (E 70323) shows the rock to be quartzitic conglomerate, composed of coarse-grained sand to small pebble-size detrital grains, dominantly of quartz, metaquartzite, cherts and opaque grains. There is a compacted quartzite texture, with complex grain boundaries, including triple point and sutured contacts. Silica authigenesis is ubiquitous as overgrowths and cements. Iron oxide or hydroxide (possibly haematite) occurs as intergranular secondary deposits.

Tunbridge (1986) studied the heavy minerals of the Trentishoe Member in and west of the district and found that zircon and tourmaline dominate, with minor rutile. Garnet was rare and usually highly etched. In contrast, samples from the highest Lower Old Red Sandstone of South Wales and the Welsh Borderlands contained garnet as the dominant heavy mineral, with zircon and tourmaline, and minor rutile. The difference in abundance of garnet between the two outcrops was explained by diagenetic dissolution of the garnet in the Trentishoe Member.

## Biostratigraphy

The Hangman Sandstone is generally barren of fossils; Tunbridge (1978) noted that the Trentishoe Member, at unspecified localities, contains scattered remains of *Psilophyton* sp. and a calamitid-like stem. In the district, fragmentary plant remains identified as *Hoftimella* sp. by Dr Chaloner (RHBNC, London) have been recorded from excavations [SS 968 453] for Hopcott service reservoir. Professor Dianne Edwards (University of Wales College of Cardiff) has examined a specimen (Plate 17) collected from Henner's Combe [SS 9141 4915] by Mr and Mrs Rodber of Minehead, and comments as follows: 'The object is certainly part of the vegetative (sterile) axis of a plant, and from the horizontal ribbing and branching pattern is probably a member of the Cladoxylaces. The best known Devonian representative is *Pseudosporochnus nodosus*, which has a Middle to basal Upper Devonian range, but it would be impossible to identify with certainty from a fragment.'

Randomly scattered plant stems occur in a thin (0.02 m) bed of greenish grey, micaceous siltstone in loose material at Culver Cliff, Minehead [SS 9614 4783] (Jones, 1995). The purple-grey stems are flat-lying, branched, typically 0.3 to 0.5 cm wide and up to 10 cm long; they show narrow striations parallel to their long axis.

Evans (1922) recorded plant remains and a scale of *Coccosteus* from the Rawn's Member in the Sherry Combe area near Combe Martin, in the Ilfracombe district. The Hollowbrook, Sherrycombe and Little Hangman members, not known to occur within the Minehead district, have locally yielded fossils in the Ilfracombe district (Tunbridge, 1978). The Hollowbrook Member is poorly fossiliferous apart from the trace fossil *Arenicolites*, although tentaculitids have been recorded. The Sherrycombe Member contains *Natica*, *Myalina* and, very rarely, fragments of *Thamnopora* in thin shelly horizons. The marine trace fossils *Chondrites* and, in the higher beds, *Diplocraterion* were also recorded.

Knight (1990) noted that the bulk of the Hangman Sandstone is barren of palynomorphs, but that palyniferous strata towards the top of the formation yielded assemblages of late Eifelian/earliest Givetian aspect.

Jones (1995) recorded the nonmarine trace fossils *Beaconites* and *Planolites* in desiccated and remobilised mudstones (Facies 7) of the Hangman Sandstone (see above).

The stratigraphical position of the Hangman Sandstone, between the Lynton Formation below and the Ilfracombe Slates above, indicates that it is largely if not entirely of Mid Devonian, probably mainly Eifelian, age, with some Givetian strata in the upper part.

## Depositional environments

Tunbridge (1984) suggested that the sequence represented an ephemeral stream/clay playa complex. However, the abundance of both sheet and channel sandstones suggests proximity to a sediment source, indicating that deposition was in a distal alluvial fan environment in which ephemeral continental mudflats developed at intervals. The sediment was supplied from the more proximal parts of the alluvial fan during periods of high rainfall.

**Plate 15**  Coastal exposures of the Hangman Sandstone: Glenthorne Beach to The Caves [SS 8019 4948 to 8042 4945] (GS495).

**Plate 16** Coastal exposures of the Hangman Sandstone: general view of Giant's Rib [SS 7948 4990] (GS496).

**Plate 17** Plant fossil (?*Pseudosporochnus nodosus*) from the Hangman Sandstone, Henner's Combe [SS 9141 4915]. Photograph courtesy of Mr H Prudden. (GS497).

Tunbridge (1984) recognised three types of facies sequences within the Hangman Sandstone, which he believed represented the transition from proximal to distal parts of an ephemeral stream/clay playa complex. The proximal sequence comprised dominantly multistorey channel sandbodies, the medial sequence consisted of sheet sandstones and channel fills, and the distal sequence comprised a clay playa facies dominated by desiccated and remobilised mudstones. Tunbridge suggested that these facies sequences represented progressive downslope changes in sedimentation. Examination of the Hangman Sandstone during the course of the survey of the district suggests that this tripartite division is likely to be valid, but that downslope changes are unlikely to be the only possible control on its development. Changes in sediment flux, fan retrogradation, and changes from an axial to lateral fan position may create similar facies sequences.

The eight facies recognised within the Hangman Sandstone, described above, include channel, sheetflood and ephemeral lake deposits. The channels were probably of low sinuosity and may have been distributive. They were filled predominantly by bedforms generated during the flood stage, including low-angle barforms, plane beds and some dunes. Waning flows are represented by the dumping of sediment to form mudstone drapes. The channels were shallow (less than 3 m deep) and were probably wide. The multistorey channels may have been formed either by single channels relocating in the same area over time, or by the action of multichannel rivers with a hierarchy of channels. Channels stack as a result of either low aggradational rates or of high sediment flux relative to rates of lateral migration and frequency of avulsion (Bridge, 1985; Bridge and Leeder, 1979). In the case of the Hangman Sandstone sequence, channel stacking is thought to be related to periods of increased sediment supply, as indicated by the evidence for rapid cutting and filling of channels.

Sheet sands formed by sheetflood events are a common feature of the Hangman Sandstone. A sheetflood is an unconfined, sediment-laden flood of water of relatively low frequency and high magnitude (McGee, 1897; Hogg, 1982). A decrease in the gradient and expansion of the flow leads to deceleration and deposition as a sheet of sand. Sheet sands in the Hangman Sandstone were deposited on a low-relief sediment surface that varied from sand- to mud-dominated. Periods of high and frequent discharge resulted in the formation of a sandflat, whereas periods of lower discharge formed a mudflat/playa lake.

Palaeocurrent measurements made by Tunbridge (1984) indicated sediment transport directions from between north and north-west, and are shown on Figure 13. Tunbridge (1981, 1984) suggested a source area in South Wales, possibly from the Middle Devonian Brownstones (highest Lower Old Red Sandstone). Measurements made during the present survey (Figure 14) confirm those of Tunbridge (1984).

The mudflat/playa lake deposits include siltstones and claystones, interbedded with sheetflood sandstones. The playas probably developed subsequent to sheetflood deposition, with pools or small lakes as ephemeral features. High rates of evapo-transpiration led to the development of small pedogenic carbonate nodules in rare instances. The groundwaters were evidently carbonate-rich rather than sulphate-rich, as indicated by the lack of evaporite minerals. A semi-arid climate is indicated by the incipient calcrete palaeosols and the ephemeral nature of sedimentation.

## Coastal exposures of the Hangman Sandstone

The following are coastal localities where the main features of the Hangman Sandstone are well seen; locations are shown on Figure 13. Further details of coastal and inland exposures of the formation are given by Edwards (1996). The only access to the 7 km stretch of coast between Glenthorne and Porlock Weir is by private track via Glenthorne, by private track to Embelle Wood via Broomstreet Farm, or westwards from Porlock Weir. There is no easy access to the 7 km section of coastline between Hurlstone Point and Minehead, except westwards from Minehead. The whole isolated coastline is difficult to access and **potentially hazardous** owing to the long distances between access points, the rocky and slippery nature of the foreshore, and the high tidal range which means that excursions have to be carefully timed.

The **Glenthorne** area, close to the western edge of the district, is a Site of Special Scientific Interest (SSSI) for the Hangman Sandstone, extending from [SS 794 499] near Giant's Rib, to [SS 805 495] near The Caves. At **Giant's Rib** [SS 7948 4990] grey-purple sheet sandstones and siltstones (mainly Facies 4), in beds up to 1.6 m thick, form prominent slabs dipping seaward (Plate 16). Desiccation cracks are common throughout the succession hereabouts. About 100 m east of Giant's Rib, well-developed convolute lamination is present in the upper part of a possible channel sandstone, 2.5 m thick, which

forms part of a 4 m-thick multistorey sandstone unit (Facies 2), underlain by sheet sandstones (Facies 4). Convolute lamination affects sandstone channel units up to 2 m thick, with dewatering pipes present in the upper 1 m of sandstone. Thick sandstone beds are common along this part of the coast, and are dominated by low- to moderate-angle cross-bedding and convolute lamination. The sheet sandstones locally contain very low-angle and open trough cross beds. Siltstones, draping undulose sandstone sheets, occur in units up to 3 m thick, always in association with thickly laminated to thinly bedded sheet sandstones and desiccation cracks. The siltstones generally lack burrows and are commonly brecciated.

At **Yellow Hammer Rock** [SS 7980 4975], near Glenthorne House, a 12 m measured sequence consists mainly of single-storey channel sandstones with some thick sheet sandstones; an interbedded/interlaminated siltstone/sandstone sequence is 2.6 m thick (Figure 15).

From **Glenthorne Beach** to **The Caves**, cliffs up to about 20 m high show a 250 m section in folded reddened sandstones, most of which is possibly Facies 4 (Plate 15). About 50 m east of Glenthorne Beach, cliffs [SS 8024 4948] up to about 12 m high are mainly in grey, thickly bedded (locally convolute-bedded), quartzitic, fine- to medium-grained sandstones in beds up to about 1.2 m thick, folded into a monocline, the axial surface of which dips south at 40°. At **The Caves** [SS 8031 4947], several narrow caves have been eroded along interbeds of siltstone and shale within a sequence of near-vertical, thickly bedded, reddened sandstones forming the steep limb of an anticline.

The sequence around **Yellow Stone** [SS 8195 4911], dips south at low angles; two measured sections are shown in Figure 16. Thick convolute-bedded channel sandstones are succeeded upwards by sheet sandstones. A channel sandstone at this locality is fine to medium grained, greenish grey, cross-bedded, and is a minimum of 2.1 m thick.

At the **Ivy Stone** [SS 8387 4876], bedding planes in sandstones dip steeply seaward. A minor, laterally accreting, 0.84 m-thick, channel sandstone is present (Plate 7). Convolute lamination is also present. A sedimentological log through a succession of desiccated sandstones and siltstones is given in Figure 17. About 350 m east-south-east of the Ivy Stone, the cliff [SS 8420 4864] shows about 20 m of grey sandstone with, in the lower half, distinctive banded and mottled, purple and pale grey-green, laminated and thinly bedded mudstone and very fine-grained sandstone. A prominent 0.3 m bed of pale grey sandstone contains large angular clasts of purple shale and siltstone.

Cliffs [SS 8436 4862] just west of Culbone Rocks show a fairly uniformly dipping sequence of up to 20 m of grey, quartzitic, locally quartz-veined, fine- to medium-grained sandstone, mainly thickly bedded, with some interbeds up to 1.8 m thick of purple siltstone. Some sandstone beds have loaded bases. At **Culbone Rocks** [SS 8442 4862], grey-purple, parallel-laminated and convolute-bedded quartzitic sandstones are locally intensely fractured.

The sequence around **First Rocks** [SS 8520 4850], about 1.5 km west-north-west of Porlock Weir, is dominated by

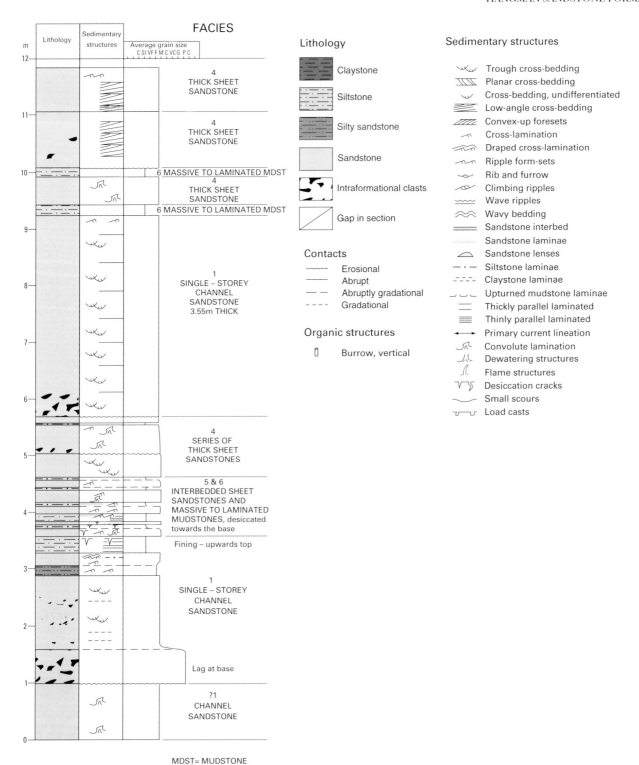

**Figure 15** Graphic sedimentary log and interpretation of the Hangman Sandstone at Yellow Hammer Rock [SS 7980 4975]. See text for full description of numbered facies.

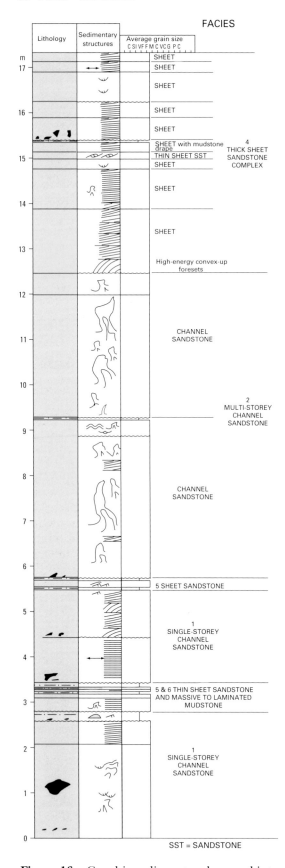

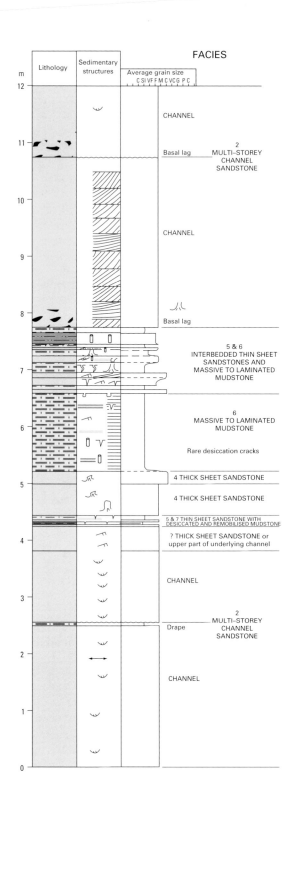

**Figure 16**  Graphic sedimentary logs and interpretation of the Hangman Sandstone at Yellow Stone [SS 8195 4911]. For key see Figure 15; see text for full description of numbered facies.

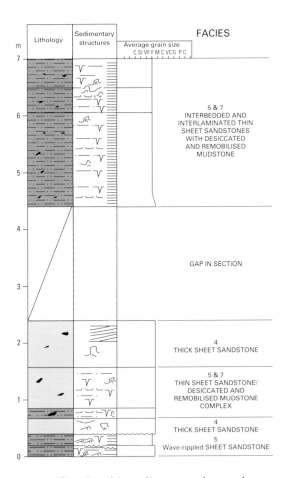

**Figure 17** Graphic sedimentary log and interpretation of the Hangman Sandstone at Ivy Stone [SS 8387 4876]. For key see Figure 15; see text for full description of numbered facies.

reddish purple siltstones and claystones with abundant desiccation cracks, with thin to thick interlaminated sandstones (Facies 5); thicker sheet sandstones (Facies 4) and rare channels are also present. Carbonate nodules are present in some mudstone beds (Plate 13).

In the southernmost cliff exposures [SS 8993 4913] at **Hurlstone Point**, the succession consists of interbedded thinly to thickly bedded, red, sheet sandstones and siltstones, with rare thin to thick claystone laminae (Facies 4) (Plate 8). Sandstones form up to 60 per cent of the succession and are variable in thickness, up to a maximum of approximately 1 m, although the thicker sandstones are generally composite in nature, as indicated by the presence of impersistent claystone laminae which in places drape ripple form sets and are commonly desiccated.

Cliffs [SS 8991 4916] at the northern end of the shingle ridge south of Hurlstone Point show grey and grey-purple, locally thickly bedded, quartzitic, fine- to medium-grained sandstones, with some interbeds of cleaved siltstone and shale, folded into a syncline (Plate 24). In the small headland [SS 8985 4918] west of the fold, multistorey channel sandstones (Facies 2) are exposed (Plate 6).

On the west-facing part of Hurlstone Point, accessible only at low tides, the sequence consists of sheet sandstones and siltstones (possibly Facies 4). A prominent set of faults [SS 8987 4925] dipping at 60° to 175° intersects the headland.

At **Minehead Bluff** [SS 9146 4937] cliffs up to 15 m high show mainly thickly bedded sandstones with some very thick convolute-bedded units (possibly Facies 1). Scattered, rounded to well rounded, white and pink quartz pebbles are present locally in grey sandstones.

In the **Greenaleigh** area, the Hangman Sandstone forms a series of isolated exposures. At a locality [SS 9440 4850] about 320 m east of Grexy Combe, a siltstone sequence, 3.2 m thick (Facies 7), is overlain by an erosively based channel sandstone 2.8 m thick (Facies 1). The sequence comprises pale reddish brown siltstone with numerous very fine- to fine-grained sandstone beds and laminae (1.2 m) with desiccation cracks, disrupted, upward-broken (dewatering-induced) mudstone laminae, and bioturbation. These are overlain by siltstone (2 m) with common carbonate nodules forming an incipient calcrete unit. Dykes filled with grey, fine-grained sandstone are locally present.

Near Greenaleigh [SS 9511 4818], thickly bedded (up to 0.9 m), locally quartz-veined sandstones, parallel-laminated in some beds, and with interbeds of cleaved shale, are seen in an anticline (Plate 25).

At the western end of **Culver Cliff**, near Minehead, the cliffs [SS 9609 4784–9617 4782] are 20 to 30 m high and consist of sheet sandstones (mainly Facies 4), with some channel sandstones and mudstones (Facies 1). The sheet sandstones are greenish grey and siliceous, thickly bedded, and vary from fine to medium grained, up to coarse grained. Mudstone rip-up clasts occur typically at the bases of sandstone beds, but also within beds. Rare, small, subangular to subrounded pebbles of quartz occur in some beds. At beach level, at [SS 9614 4783], a distinctive 3.2 m-thick bed of grey quartzitic sandstone contains large synsedimentary (slump) folds (Plate 5). The major structural feature of Culver Cliff is a monocline [SS 9614 4783] with a fractured axial zone (Plate 26).

## Mineralisation

At Combeshead Quarry [SS 9283 4767], beds of purple siltstone and fine sandstone dipping at 65° to 317° are cut by joints and thin veins filled with pink barite, with margins coated with dark red haematite. In the central part of the north face of the quarry, a brecciated vein up to 0.2 m wide is cemented by barite and haematite. The veins and joints strike 321° and dip eastwards at 75°, suggesting formation during a period of east–west extension. A thin section (E 71335) of a 35 mm-wide barite veinlet shows a 4 mm-wide selvedge of earthy and specular haematite overgrown by 5 mm of haematite and barite; the core of the veinlet is a felted mass of colourless barite crystals with abundant small inclusions of haematite. The barite crystals in places contain small (less than 15 $\mu$m) liquid-filled inclusions, mostly of irregular shape. The absence of vapour bubbles in these inclusions suggests that the barite was deposited at low temperatures (probably less than 120°C).

This mineralisation appears to be genetically distinct from the nearby Brendon Hills iron veins which are of east–west trend and comprise veins of siderite, much altered to limonite and related species in the near-surface zone of oxidisation and weathering (Dines, 1956). Fluid inclusion and other data quoted by Dr U F Hein of the Institute for Geology at Göttingen (1995) suggest that the Brendon Hills mineralising fluids were NaCl brines of low/moderate salinity with minimum trapping temperatures in the range 210°C to 330°C. Liquid $CO_2$ is present in some of the inclusions. Hein concludes that these Exmoor ore fluids were of metamorphic origin, and formed during the late stages of the Variscan orogeny. A later event is suggested for the Combeshead Quarry barite veins, which are possibly related to the Triassic lead-zinc-barite-fluorite 'crosscourse' mineralisation described by Scrivener et al. (1994) elsewhere in the south-west England metallogenic province.

# FIVE

# Permo-Triassic

During the post-orogenic extensional phase that followed the formation of the Variscan foldbelt in south-west England, a sequence of mainly continental red beds was deposited unconformably on folded Devonian and Carboniferous rocks. The red beds in the Minehead district are included in the New Red Sandstone Supergroup (Laming, 1968) and are probably mainly Triassic in age. They are overlain by the Penarth Group, a predominantly grey-coloured sequence, mainly of marine origin.

The New Red Sandstone Supergroup comprises material eroded from the Variscan mountains and laid down initially on an irregular land surface, and in fault-controlled basins. Elsewhere in south-west England, the youngest rocks affected by the Variscan Orogeny are of Late Carboniferous (Bolsovian: Westphalian C) age (Ramsbottom et al., 1978); in south Devon, the oldest New Red Sandstone is of Early Permian and possibly even latest Stephanian (Late Carboniferous) age (Edwards et al., 1997). In the Minehead district, removal of any Carboniferous rocks present, and exposure of a major part of the Devonian sequence, had taken place before the first breccias of the New Red Sandstone Supergroup (Luccombe Breccia Formation) began to accumulate in the Porlock Basin. The age of these unfossiliferous breccias remains uncertain; they are thought to be Triassic, but a Permian, or even latest Carboniferous (Stephanian), age for the lowest beds cannot be discounted. The upper part of the New Red Sandstone of the district, and the succeeding Penarth Group, are dateable by fossils, as Triassic. The top of the Penarth Group does not coincide with the Triassic–Jurassic boundary, which occurs a few metres above the base of the succeeding Lias Group (Chapter 6). The lowest beds of the Lias Group are, therefore, of Triassic age, but are described, for convenience, with the overlying (Jurassic) part of that group (Chapter 6).

The term 'Permo-Triassic' is used in this memoir for the sequence between the base of the New Red Sandstone and the base of the Jurassic, but, as noted above, it carries with it an implicit uncertainty regarding the presence of Permian strata.

In the district, the Permo-Triassic rocks are represented both onshore and offshore by up to about 1100 m of breccias, conglomerates, sandstones and mudstones, which rest with marked unconformity on Devonian rocks, and pass up conformably into Jurassic rocks. Onshore, owing to their lesser resistance to erosion, the Permo-Triassic rocks form lowland areas bordered by hills of more resistant Devonian sandstones. The distribution of the lithostratigraphical units in the Permo-Triassic (Table 6) is shown on Figure 18.

The earliest New Red Sandstone deposition in the district occurred in separate, north-west-trending basins, the Porlock Basin in the west and the Minehead Basin in the east (Figure 18). These basins are half-grabens, defined by major faults on their north sides and largely unfaulted on the south; the Porlock Basin is also partly bounded by major north–south faults. Rocks infilling the basins dip to the north so that successively younger beds crop out from south to north. Owing to the separation of the Porlock and Minehead basins, different local sequences formed in each until mid-Triassic times (during the deposition of the Mercia Mudstone Group), when more uniform conditions were established over the whole district.

The Mercia Mudstone and Penarth groups occur in both the Porlock and Minehead basins; the deposits there are contiguous with representatives of those groups present at subcrop in most of the offshore area. In the Minehead Basin, the Sherwood Sandstone Group is represented by the Budleigh Salterton Pebble Beds and Otter Sandstone Formation. In the Porlock Basin, however, the Mercia

**Table 6** Lithostratigraphical classification of the Permo-Triassic rocks of the district.

| Group | Porlock Basin | Minehead Basin–Watchet | Group |
|---|---|---|---|
| Penarth Group | Lilstock Formation*<br><br>Westbury Formation* | Lilstock Formation*<br><br>Westbury Formation* | Penarth Group |
| Mercia Mudstone Group | Blue Anchor Formation<br><br>Undifferentiated (red Mercia Mudstone) | Blue Anchor Formation<br><br>Undifferentiated (red Mercia Mudstone) | Mercia Mudstone Group |
| | Luccombe Breccia Formation | Otter Sandstone Formation | Sherwood Sandstone Group |
| | | Budleigh Salterton Pebble Beds | |

\* not shown separately on the 1:50 000 scale map

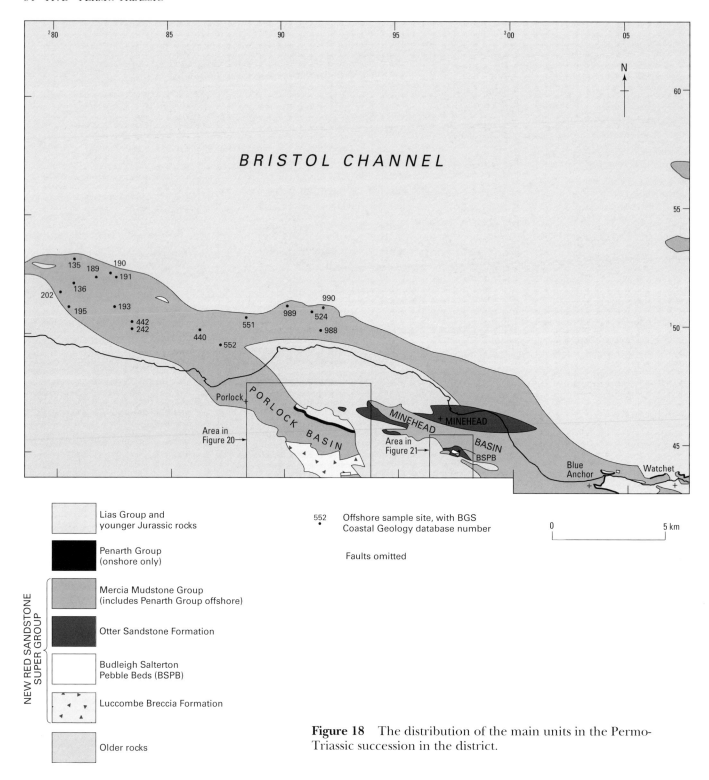

**Figure 18**   The distribution of the main units in the Permo-Triassic succession in the district.

Mudstone is underlain by the Luccombe Breccia which occurs only in that basin; its correlation with pre-Mercia Mudstone sequences elsewhere is therefore uncertain. The generalised Permo-Triassic and Jurassic sequences in the Porlock and Minehead basins are shown on Figure 19.

The relationship of the basal New Red Sandstone Supergroup to the underlying Devonian 'basement' varies in different parts of the district. In the Porlock Basin, the Luccombe Breccia locally forms the base of the sequence on the south side of the basin, while in the south of the Minehead Basin, around Alcombe [SS 975 450] and between Bratton [SS 953 460] and Hindon Farm [SS 933 467], it is the Budleigh Salterton Pebble Beds and the Otter Sandstone which rest on the Devonian rocks. However, all these formations are in places overlapped by the Mercia Mudstone so that along much of the southern

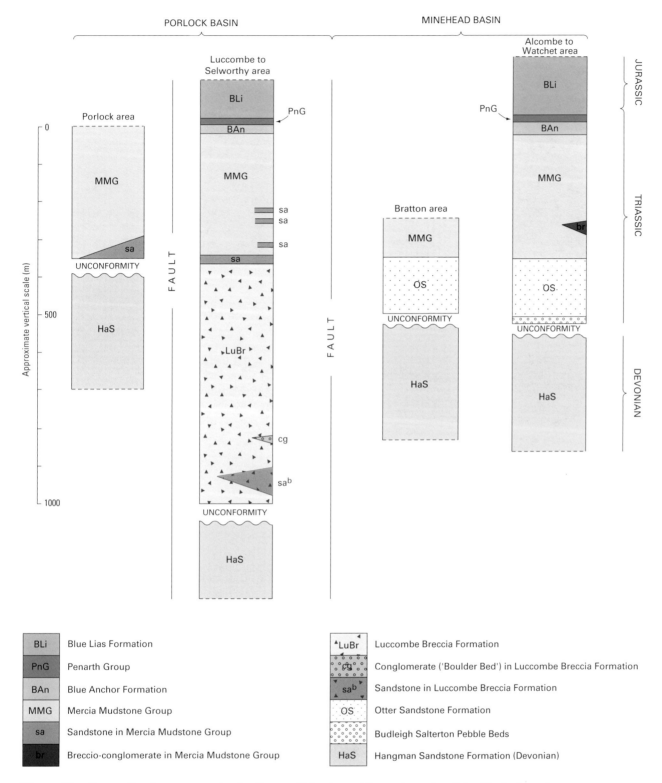

| | |
|---|---|
| BLi | Blue Lias Formation |
| PnG | Penarth Group |
| BAn | Blue Anchor Formation |
| MMG | Mercia Mudstone Group |
| sa | Sandstone in Mercia Mudstone Group |
| br | Breccio-conglomerate in Mercia Mudstone Group |

| | |
|---|---|
| LuBr | Luccombe Breccia Formation |
| cg | Conglomerate ('Boulder Bed') in Luccombe Breccia Formation |
| sa^b | Sandstone in Luccombe Breccia Formation |
| OS | Otter Sandstone Formation |
| | Budleigh Salterton Pebble Beds |
| HaS | Hangman Sandstone Formation (Devonian) |

**Figure 19** Generalised sequences in the Permo-Triassic and Jurassic rocks of the Porlock and Minehead basins.

side of the Minehead Basin between Dunster and Periton, and west of Horner in the Porlock Basin, that group forms the base of the New Red Sandstone sequence and rests upon the Devonian. In the Dunster area, hills of Devonian sandstone protruding from the surrounding Mercia Mudstone represent an exhumed Triassic topography (Thomas, 1940). Coarse marginal deposits of the Mercia Mudstone are not widely developed in the district, in contrast to areas north of the Bristol Channel and in the Mendips, where such deposits are common. However, around Dunster, there are local developments of breccio-conglomerates, breccias and sandstones which are probably marginal Mercia Mudstone deposits (p.70). East of Dunster, the basal unconformity of the New Red Sandstone is not seen again in the district; it is visible about 4 km east of the district, in the West Quantoxhead area, where Whittaker and Green (1983) recorded a partially exhumed Triassic topography, with red mudstones resting on the Devonian Hangman Sandstone, marginal deposits being apparently absent.

The Penarth Group, the Blue Anchor Formation, and the upper part of the red Mercia Mudstone can be satisfactorily correlated between the Porlock Basin and the Minehead Basin, on the basis of lithology and biostratigraphy. Correlations of older New Red Sandstone Supergroup strata are hindered by the lack of reliable dating evidence, and the possibility of diachronous relationships. The Luccombe Breccia lies below the Mercia Mudstone in a position analogous to that of the Otter Sandstone and may thus be, in part, of similar Triassic age. However, in the absence of good exposures, it is not known with certainty whether there is a major break between the Mercia Mudstone and Luccombe Breccia; a discrepancy in the dips of the two units (p.68) suggests the possibility of angular discordance between them. Thomas (1940) considered that the upper part of the Luccombe Breccia was of 'lower "Keuper" age', that is equivalent to the Otter Sandstone farther east. He also thought it probable, in view of the great thickness of the breccia, that the lower part is the same age as the '"Bunter" Pebble Beds' (now the Budleigh Salterton Pebble Beds in south-west England); this suggestion implied that the Pebble Beds and a bed of similar coarseness (the 'Luccombe Boulder Bed' (p.59)), within the Luccombe Breccia, are correlatives. There is, however, no firm evidence for this hypothesis, and the lack of clasts of Carboniferous Limestone in the 'Luccombe Boulder Bed' is an argument against it.

Correlation of the sandstones of the supergroup between the two basins is uncertain. In the Minehead Basin, the attribution of sandstones at Alcombe to the Otter Sandstone is clear from their stratigraphical position between the inferred Budleigh Salterton Pebble Beds and the Mercia Mudstone. The sandstones in the western part of the Minehead Basin, between Bratton and Hindon Farm, apparently underlie Mercia Mudstone and are here also classified as Otter Sandstone. However, Thomas (1940) thought it probable that the sandstones in the latter area are a marginal facies of the Mercia Mudstone. The stratigraphical position of sandstones, breccias and minor conglomerates which crop out in the

older part of Minehead town is uncertain, owing to their outcrop being wholly fault bounded, and because of lithological differences from the Otter Sandstone at Alcombe. They are included here in the Otter Sandstone, but differ from the mottled Otter Sandstone of Alcombe in consisting of cross-bedded sandstones with units of breccia and minor conglomerate.

In the Porlock Basin, sandstones lying between the Luccombe Breccia and the Mercia Mudstone were termed '"Keuper" Sandy Limestones' by Thomas (1940); he considered that they were of the same age as the sandstones west of Minehead.

Biostratigraphical evidence indicates that the Penarth Group and the upper part of the Mercia Mudstone are Triassic in age. In the Porlock Basin, the lowest productive palynology sample, from about 75 m above the base of the Mercia Mudstone, yielded miospores indicative of a Carnian to Rhaetian (Late Triassic) age (p.73). The Luccombe Breccia in the Porlock Basin underlies the Mercia Mudstone and is pre-Carnian. Its lower age limit is constrained by the age of the youngest rocks involved in the Variscan Orogeny, which elsewhere in south-west England are Late Carboniferous (Bolsovian); thus, the age of the Luccombe Breccia is between Bolsovian and Carnian. In the Minehead Basin, the oldest formation of the New Red Sandstone is the Budleigh Salterton Pebble Beds, believed, from evidence outside the district, to be of probable Early Triassic (Induan–Olenekian) age. There is no direct evidence of the presence of Permian rocks in the district.

The Permo-Triassic sequence of the district records the upward change from a largely continental environment to marine conditions. Nonmarine conditions were brought to an end by a marine transgression from the Tethyan province to the south in latest Triassic times. The first indications of the transgression are in the uppermost beds of the Blue Anchor Formation of the Mercia Mudstone, and its progress is recorded in the succeeding dominantly marine Penarth Group. By Lias Group times, fully marine conditions had been established, and a shallow epicontinental sea extended over much of southern Britain.

## LUCCOMBE BRECCIA FORMATION

The Luccombe Breccia Formation ('Luccombe Breccia Series' of Thomas, 1940) is the lowest unit of the New Red Sandstone Supergroup succession in the Porlock Basin. Its outcrop extends from near Horner in the west, through Luccombe, to Huntscott and Tivington Knowle in the east (Figure 20). To the south, the formation rests unconformably on Hangman Sandstone (Devonian). To the north, it is overlain, possibly unconformably (p.68), by the Mercia Mudstone. To the west, the outcrop is bounded by the north–south Horner Fault; west of this fault, the Luccombe Breccia is missing, and the Mercia Mudstone (with basal sandstone) rests directly on Devonian rocks. To the east, the Luccombe Breccia is faulted against Devonian rocks by the major north-south Tivington Fault which is offset by an east-west fault

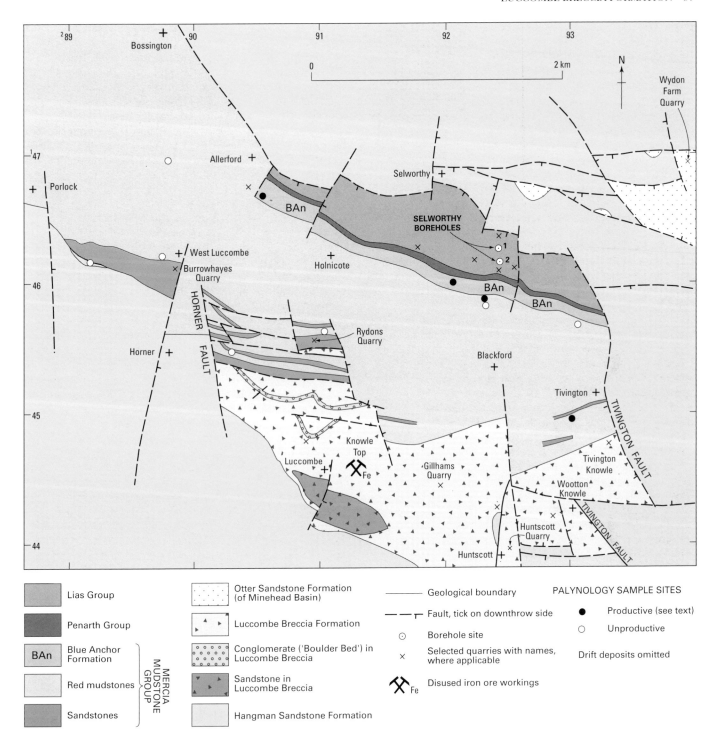

**Figure 20** Geological map of part of the Porlock Basin. See Figure 18 for location.

LEGEND:

- Lias Group
- Penarth Group
- BAn — Blue Anchor Formation
- Red mudstones — MERCIA MUDSTONE GROUP
- Sandstones — MERCIA MUDSTONE GROUP
- Otter Sandstone Formation (of Minehead Basin)
- Luccombe Breccia Formation
- Conglomerate ('Boulder Bed') in Luccombe Breccia
- Sandstone in Luccombe Breccia
- Hangman Sandstone Formation

- Geological boundary
- Fault, tick on downthrow side
- ⊙ Borehole site
- × Selected quarries with names, where applicable
- Fe Disused iron ore workings

PALYNOLOGY SAMPLE SITES
- ● Productive (see text)
- ○ Unproductive

Drift deposits omitted

through Wootton Knowle (Figure 20). South of the district, east of a fault through Huntscott, the breccias are again absent and a thin unit of sandstone beneath the Mercia Mudstone rests directly on Devonian rocks (Thomas, 1940). Possibly, there was a period of faulting after deposition of the Luccombe Breccia, followed by erosion (of uncertain duration) before deposition of the Mercia Mudstone Group. This interpretation is sup-

ported by the possible presence of an angular unconformity between the Luccombe Breccia and the Mercia Mudstone (p.68).

The Luccombe Breccia consists predominantly of reddish brown breccias, with local interbeds of sandstone and pebbly sandstone (Plate 19). Sandstone is locally dominant over breccia. All the lithologies are commonly very calcareous, calcite-veined in places, and locally well cemented. The calcium carbonate contents of seven analysed samples range between 28.5 and 54.1 per cent (average 38.4 per cent). Calcite veining is particularly prominent at, for example, Gillhams Quarry [SS 9197 4445], where a banded calcite vein about 3.5 m thick is present (p.61).

South of Luccombe, a lenticular unit of reddish brown, weakly cemented, fine-grained sandstone has been mapped near the base of the formation. North of Luccombe, and extending westwards towards Horner, a distinctive bed of conglomerate (the 'Luccombe Boulder Bed' of Thomas, 1940) also forms a separate mappable unit. Thomas (1940) distinguished and mapped 'shaly breccias' and 'coarse breccias' within the main mass of the Luccombe Breccia; between Horner and Huntscott, the coarse breccias were shown as a unit about 12 m thick beneath the 'Luccombe Boulder Bed', thickening in the Wootton Knowle and Tivington Knowle areas to occupy much of the thickness of the formation. However, the results of this survey indicate that there is no satisfactory evidence to subdivide the breccias within the formation on the basis of clast size.

The breccias give rise to a distinctive topography of small steep-sided hills ('knolls') which contrast with the more subdued and partly drift-covered ground to the north, underlain by the Mercia Mudstone. Small disused quarries scattered over the Luccombe Breccia outcrop were formerly worked for building stone and lime, and at Knowle Top [SS 913 445], near Luccombe, old workings for iron ore in mineralised breccias are present (pp.16, 60).

The breccias in the formation dip northwards at between 30 and 54°, the average of 15 dips being 40°. Cross-sections based on the measured dips suggest that the Luccombe Breccia has a maximum thickness of up to 650 m. Thomas (1940, p.27) suggested a considerably smaller thickness of up to 1006 feet (306 m) at Luccombe.

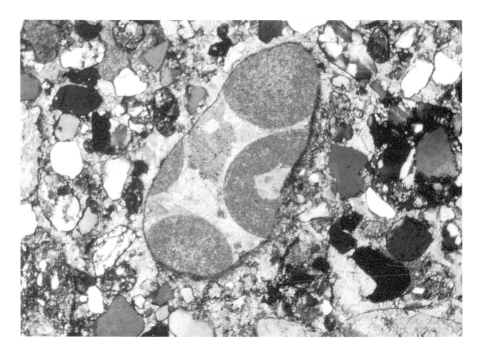

**Plate 18**   Photomicrograph ($\times$ 43) of thin section of calcareous sandstone in the Luccombe Breccia at Huntscott [SS 9257 4394], showing clasts of ooidal limestone (E 70299) (GS498).

**Plate 19**   Luccombe Breccia, south side of Gillhams Quarry [SS 9206 4441]; reddish brown, calcite-veined, sandy breccias with interbeds of sandstone. The hammer is 0.3 m long (GS499).

**Plate 20** 'Boulder Bed' in the Luccombe Breccia at Huish Ball Steep [SS 9119 4503], Luccombe. The hammer is 0.3 m long (GS500).

The breccias are poorly sorted, clast-supported, and have a fine-grained sandstone matrix. They contain platy, low-sphericity, angular to subangular shale or slate clasts. Slate clasts are typically dark grey to dark reddish grey. Other clasts include fine-grained sandstones, vein quartz, and a few laminated, reddish brown to pale creamy yellow siltstones. The clasts are generally similar in size, typically medium to large pebbles, less commonly up to very large pebble size. Beds are typically 0.2 to 0.5 m thick and form sheets with erosive bases and sharp tops; they commonly show coarse-tail grading, characterised by a decrease in the size of clasts, although the matrix typically remains constant in grain size. Rare imbrication occurs. Rare to common, coarse to very coarse, well-rounded, frosted grains occur in the matrix. Beds of sandstone, forming part of fining-upward successions, are fine to medium grained, pebbly and poorly sorted.

The erosive bases and fining-upward nature of the breccias indicate that tractional processes were dominant during their formation, and that they were probably deposited by unconfined coarse sheetfloods. The orientation of clasts indicates that they were probably carried in a flat-lying form within the flow; alternatively, they may

have been imbricated slightly and rotated to a near horizontal position by later compaction. The presence of well-rounded, frosted grains indicates that sediment in the source area underwent reworking by aeolian processes before subaqueous transportation.

Around Huntscott, the sequence is dominantly of calcareous pebbly sandstones, sandstones and sandy breccias, which are commonly well-cemented. Calcite veins are locally abundant. The pebbly sandstones predominate and probably represent the deposits of sandy sheetfloods. They are pale reddish brown, fine to medium grained, and contain abundant well-rounded, frosted grains. Extraformational clasts are present, and are typically concentrated towards the bases of beds; they are generally small to large pebble-grade, angular, of low-sphericity, and are flat-lying. Most clasts are dark purple-grey, fine-grained limestone; rare vein quartz is present.

Other sandstones at Huntscott are probably aeolian in origin. They are pale reddish brown, thick-bedded, cross-bedded, fine- to medium-grained sandstone. The grains are dominantly rounded, moderately to well sorted, and frosted. Cross-bedded sets are up to 0.4 m thick, with occasional very thin, purple silty laminae on the foresets, which are directed to the north (to 354°). The absence of pebbles, the lack of evidence of erosion, and the presence of abundant rounded frosted grains suggest that the sandstones originated as aeolian dunes.

Thin sections (E 70299, 70306) of the coarser sheet-flood facies rocks at Huntscott show them to be fine- to very coarse-grained calcareous sandstones, composed of rounded, very coarse-grained sand and granule-sized clasts of limestones, with mono-crystalline and polycrystalline quartz grains, cherts and minor potassium feldspar, just grain-supported in an inequigranular calcite spar cement (Plate 18). The detrital grains have a distinct ferruginous coating ('desert varnish') (Strong, 1995). The limestone clasts contain oolite, peloids and foraminifera (p.60).

A thin, apparently laterally impersistent, unit of conglomerate (the 'Luccombe Boulder Bed' of Thomas, 1940), distinguished from the main mass of the Luccombe Breccia by its coarse clast size and the degree of rounding of the clasts (Plate 20), occurs near Luccombe (Figure 20). It forms a small but well-defined ridge about 1 km long, faulted at both ends, extending from Copperclose Wood [SS 9045 4524], Horner, to [SS 9137 4504] near Huish Ball Steep, Luccombe. A few small exposures and scattered cobbles and boulders of sandstone are present along the ridge. The conglomerate outcrop is repeated by strike faulting south of the main outcrop, and forms the prominent knoll [SS 9098 4483] above East Luccombe Farm. The full thickness (5.5 m) of the bed is exposed in a lane-cutting at Huish Ball Steep [SS 9119 4503]. The conglomerate has an abrupt base, probably on an erosion surface, and fines upwards into fine-grained breccia. It is clast-supported, poorly sorted and unbedded, with no grading. The clasts are of low sphericity (but better rounded than those in the breccia) and are flat-lying, with long axes parallel to bedding in some cases; possible clast imbrication dips to the north. The matrix comprises very fine- to fine-grained sandstone, and is weakly to moderately cemented. Well-rounded medium to coarse grains, possibly ooids, are common in the matrix, particularly towards the base of the conglomerate bed. The conglomerate is oligomictic, containing almost exclusively sandstone clasts and rare vein quartz; the clasts average large cobble size, although boulders are common.

The conglomerate probably represents the deposits of a fluvial channel. Flow confinement is indicated by the sharp, probably erosive base, and the clast-supported texture indicates the operation of tractional processes during deposition. It is likely that channel depths were similar to the thickness of the deposit, that is, about 5 m. The clast composition is similar to that of the adjacent finer-grained breccias, suggesting a similar source for both units. The coarse, well-rounded grains are thought to be ooids, incorporated into the sediment by reworking of an earlier deposit (Jones, 1995).

South of Luccombe, a lenticular unit of sandstone up to 70 m thick occurs near the base of the formation (Figure 20). Exposures [SS 9093 4429 to 9097 4435] are restricted to a small stream south of Luccombe, where the sandstones are underlain by about 15 m of breccia resting with probable unconformity on Hangman Sandstone (Devonian). The sandstones are reddish brown, fine grained and moderately well sorted. Scattered granules and pebbles of sandstone and vein quartz are present along discrete layers; these clasts are flat-lying and of low sphericity. The sandstones are weakly cemented and locally weathered to sand. Interpretation is hindered by the lack of exposure, but a fluvial origin is considered most likely.

The distinctive limestone clasts (Plate 18) present in calcareous sandstones near the base of the Luccombe Breccia at Huntscott contain peloids, oolites and foraminifera. A thin section (E 70299) [SS 9257 4394] shows calcispheres and plurilocular foraminifera, including cf. *Palaeospiroplectammina* sp.; the foraminifera could be derived from Upper Devonian or Lower Carboniferous rocks. Another thin section (E 79306) [SS 9254 4395] contains *Earlandia* sp. and *Priscella* sp.; the latter indicates a Tournaisian–Namurian age (Riley, 1995).

Ooidal limestone clasts similar to those at Huntscott have been recorded from calcareous sandstones at the base of the Mercia Mudstone at West Luccombe [SS 8985 4613] (p.69) and were also noted by Thomas (1940, plate 2A) in sandstones near Hindon Farm [SS 933 467], towards the western end of the Minehead Basin.

Thomas (1940, pp.31–32) studied the heavy minerals of the Luccombe Breccia. Ilmenite was abundant, and haematite, zircon and leucoxene were common. Occurrences of fluorite, garnet and apatite were noted; barite was abundant in some samples. This mineral suite was virtually identical with that in the underlying Devonian rocks in the vicinity of the breccia outcrop.

The Luccombe Breccia was exposed in several disused quarries; details are given by Edwards (1996). Accessible (1995) exposures of the breccias are at the eastern end of Gillhams Quarry, about 1 km east of Luccombe, where shallow workings on the south side [SS 9206 4441] show 5 m of reddish brown, moderately cemented, calcite-veined, sandy breccias with interbeds of reddish brown sandstone (Plate 19). In a deep excavation [SS 9197 4445]

at the western end of this quarry, a thick (c. 3.5 m) banded calcite vein complex dips south (p.61). Typical small exposures [SS 9335 4478] of breccia are also present on the east side of a former quarry near Tivington Knowle, and show reddish brown, moderately cemented, locally calcite-veined breccia, with angular and subangular platy clasts of sandstone, slate, siltstone and vein quartz up to a maximum of 6 cm across. Locally the breccia is sandstone-rich, with interbeds of reddish brown, fine-grained sandstone. Exposures of the full thickness (5.5 m) of the 'Luccombe Boulder Bed', together with about 4 m of the underlying shaly breccias, are present in a lane cutting [SS 9119 4503] in Huish Ball Steep, Luccombe (Plate 20).

## Depositional environments

Deposition of the Luccombe Breccia on the distal parts of an alluvial fan is indicated by the dominance of sheet-flood deposits. Rare fluvial channel deposits are present, indicating a transition to alluvial plain sedimentation. Alluvial fan sedimentation is also indicated by the angular nature of many of the clasts and the immature nature of the lithologies, which are typically poorly sorted. Some influence by aeolian processes is suggested by the occurrence of well-rounded, frosted grains in the breccias and pebbly sandstones, and by the rare preservation of aeolian sandstone facies. Aeolian influence accords with a distal alluvial fan environment, which is commonly subject to ephemeral subaqueous deposition, allowing aeolian reworking and the formation of wind-blown dunes.

A local source area for the Luccombe Breccia is indicated by the clast types, which include sandstone probably derived from the Devonian Hangman Sandstone, and limestone clasts that could be derived either from nearby Dinantian limestones or from the Westleigh Limestone (Asbian–Brigantian) (Thomas, 1963 a, b). Possible ooid grains are also present in the conglomerate bed at Huish Ball Steep, requiring a source of oolitic limestone. Fan development requires both uplift of the source area and subsidence in the basin. The period of deposition of the Luccombe Breccia was probably characterised by active extension and fault-related subsidence.

## Mineralisation

Iron mineralisation is present in the Luccombe Breccia at Wychanger or Knowle Top Mine [SS 913 445] near Luccombe (Figure 20). Two open workings on the hill top trend west-north-west (p.16). The few exposures (e.g. [SS 9133 4454]) show breccia partly replaced and cemented by earthy red haematite. Dump material scattered around the eastern end of the northern open working shows examples of intense replacement of breccia by hard, purplish red haematite, cementing clasts of quartz. The haematite is mainly reniform and massive; locally it is spongeous. Veinlets of white barite occur in places in the ore, and joint surfaces are commonly coated with pyrolusite or wad. Dines (1956) recorded thin veins and small nodules of dark red, massive haematite, and thin veins of barite showing comb structure.

At the western end [SS 9197 4445] of Gillhams Quarry (Figure 20), a banded calcite vein complex some 3.5 m thick dips about 40° south and cuts across the bedding of the Luccombe Breccia which dips at 40° to the north-north-east. Within the vein complex, numerous bands of white and pink calcite enclose rafts of sandstone and breccia, commonly with striking comb structures. The rafts are much veined by calcite and commonly exhibit slickensides. At the top of the northern face of the quarry, the calcite vein material is altered to a decalcified mass of buff-coloured, earthy material with specks of wad.

Throughout, the calcite vein material shows scattered platy crystals of specular haematite, which in places form aggregates. Nests of cream-coloured carbonate, probably dolomite or ankerite, are present in places. Earthy red haematite is common on joints and on slickensided surfaces. A thin section (E 71331) of calcite vein material shows the banding to be marked by the presence of varying amounts of disseminated earthy and specular haematite, and by variation in calcite crystal size and form. Grains of quartz from the sandstone host rock are also distributed along irregular layers. In places, the calcite bears rare liquid-filled inclusions, mostly of irregular shape and up to 25 microns in diameter. Some of these inclusions carry liquid only; others contain, in addition, a small vapour bubble. Microthermometric measurements have not been undertaken, but visual inspection of fluid inclusions suggests that the calcite formed at temperatures below 150°C.

The mineralogy and textures of the vein complex, together with the fluid inclusion evidence, suggest that the deposit was precipitated from low-temperature hydrothermal fluids in an extensional tectonic regime. It is likely that the Knowle Top iron deposit was of similar low-temperature origin, though with replacement rather than fracture filling as the principal ore-forming mechanism. In this respect, it shows some similarity with the manganese ores of the Crediton Trough in the Exeter district to the south (Edwards and Scrivener, in press), where manganese oxide and carbonate minerals selectively replace Permian breccia and sandstone. The fluids responsible for this type of mineralisation are considered to have been basinal brines formed in red beds in Permian and Triassic sedimentary basins and expelled during early Mesozoic extensional tectonic events (Scrivener et al., 1993).

No detailed geochemical analyses are available for the Knowle Top Mine or Gillhams Quarry deposits, but one sample from the former site and three from the latter were analysed for gold. All showed significant gold enrichment, with an averaged value of 23 parts per billion (ppb) for Gillhams Quarry, and 20 ppb for Knowle Top Mine, compared with the average value for crustal rocks of 4 ppb (Levinson, 1974).

## BUDLEIGH SALTERTON PEBBLE BEDS

In the district, beds assigned to the Budleigh Salterton Pebble Beds are preserved only in one isolated and mainly fault-bounded outcrop in the Alcombe area, on the south side of the Minehead Basin (Figure 21). The correlation of the Alcombe conglomerates with the Budleigh Salterton Pebble Beds is based on lithological similarities with undoubted outcrops of that formation in the Sampford Brett area, about 12 km east-south-east of Alcombe (Edmonds and Williams, 1985). The outcrop is 340 m long and 25 to 70 m wide. To the west and east, the formation is faulted against the Otter Sandstone. To the south it is probably in faulted contact with red mudstone which is interpreted as Mercia Mudstone, although the possibility that it is the Aylesbeare Mudstone Group, in normal stratigraphical contact with the Pebble Beds, cannot be excluded. On the north side, there is a probable normal boundary with the Otter Sandstone. Measured dips are to the north at up to 20°. The base of the formation is not exposed; it probably rests directly on Devonian rocks, but within a short distance to the east and west of Alcombe, it is overlapped by the Otter Sandstone which rests on Devonian rocks.

The formation consists of reddish brown, moderately to well-cemented, poorly sorted, clast-supported conglomerate. Bedding is generally poorly developed, but a few impersistent sandstone lenses are present locally. Clasts are of moderate sphericity, typically subrounded to rounded and of small cobble size, varying up to large cobble. They are mainly of grey and pink micritic Carboniferous Limestone, purple Hangman Sandstone (Devonian), and vein quartz, in a matrix of reddish brown, poorly sorted, fine- to medium-grained up to granular, calcareous sandstone (Plate 21). A thin section (E 70298) [SS 9762 4483] of a limestone pebble from Alcombe Quarry showed it to be a bioclastic grainstone composed of close-packed biotic debris (corals, bryozoans, gastropods, echinoderm fragments, ostracods, calcispheres, comminuted debris), with an intergranular matrix of micrite. The fauna indicates derivation from outcrops of Carboniferous Limestone (see below). Another thin section (E 70312) [SS 9772 4481] from east of Alcombe Quarry showed the Pebble Beds matrix to consist of barite-cemented medium- to coarse-grained sandstone, with minor gypsum. The presence of barite was confirmed by use of a scanning electron microscope. The detrital grains are loosely packed, some floating in the coarse sparry barite; they are dominantly quartz, with some cherts, minor feldspar including microcline, and opaque grains. Most of the detrital grains have a coating of iron oxide or hydroxide.

The beds were first described by Horner (1816, pp.358–359), who noted the occurrence in them of limestone pebbles which were picked out to be used in the manufacture of lime. The pebbles were considered by Horner to have been of Devonian age, but Martin (1909) identified Carboniferous Limestone fossils in limestone pebbles from the Budleigh Salterton Pebble Beds at Vellow and Woolston (in Sheet 295 district). A thin section (E 70298) of a limestone pebble from Alcombe Quarry [SS 9762 4483], in the Minehead district, contains the foraminifer *Earlandia* sp., and *Sphaerinvia piai*, a microfossil of uncertain affinity; the latter indicates a late Devonian (late Famennian) to early Carboniferous (early Viséan; late Chadian) age (Riley, 1995).

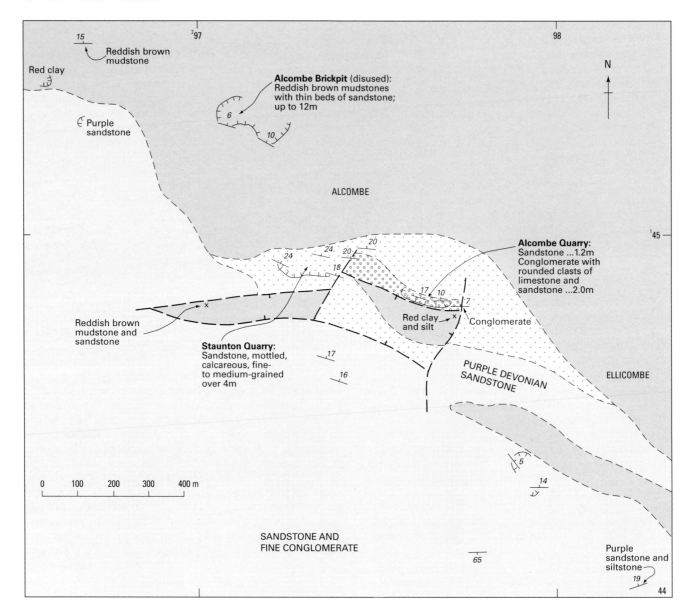

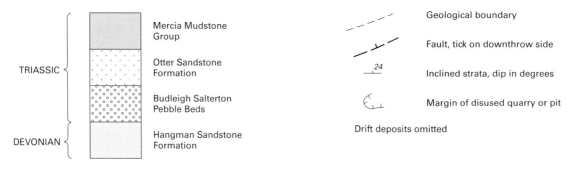

**Figure 21** Geological map of the Alcombe area. See Figure 18 for location.

**Plate 21**  Budleigh Salterton Pebble Beds at Alcombe Quarry [SS 9762 4483], showing rounded clasts of Carboniferous Limestone and Hangman Sandstone. The hammer is 0.3 m long (GS501).

Up to 4 m of the formation are currently (1995) visible on the east side of Staunton Quarry [SS 9742 4490], but Ussher (BGS MS) recorded up to 12.2 m elsewhere in the quarry. The maximum thickness of the formation is uncertain, but is estimated to be at least 15 m.

The Budleigh Salterton Pebble Beds of Alcombe were possibly formerly in physical continuity with those of the main outcrop farther east; the present isolation of the Alcombe outcrop may be the result of faulting and erosion which resulted in the removal of most of the Pebble Beds before deposition of the Otter Sandstone, for the most part directly and unconformably on the Devonian rocks, as in the area immediately adjacent to Alcombe.

There is no direct biostratigraphical evidence of the age of the formation in the district. From evidence outside the district, an Early Triassic (Induan–Olenekian) age is usually assigned to the formation.

The formation is (1995) exposed in Staunton Quarry and Alcombe Quarry. In Staunton Quarry, the best exposure [SS 9742 4493] is a fault surface showing 3 m of reddish brown, poorly sorted, unbedded, clast-supported conglomerate with rounded, subrounded and subangular pebbles and cobbles up to 12 cm across, consisting mainly of pale grey micritic limestone and purple sandstone, in a well-cemented matrix of reddish brown, poorly sorted, calcareous, fine- to coarse-grained and granular sandstone. In Alcombe Quarry, an exposure [SS 9762 4483] shows 1.2 m of possible Otter Sandstone on 2 m of Budleigh Salterton Pebble Beds which consist of moderately cemented, poorly bedded, poorly sorted, clast-supported conglomerate. The clasts are subrounded and rounded, mainly small cobble size, varying up to large cobbles (0.2 m), of pink and grey limestone and sandstone, in a matrix of reddish brown, poorly sorted, fine- to medium-grained up to granular, calcareous sandstone, with some impersistent sandstone beds (Plate 21).

Further details are given by Edwards (1996).

**Depositional environments**

The clast-supported texture of the Budleigh Salterton Pebble Beds conglomerates indicates that tractional processes operated during their deposition. They probably accumulated in fluvial channels on an alluvial plain or distal alluvial fan. No palaeocurrent measurements were possible, and consequently the direction of flow within the channels is unknown.

South of the district, deposition of the Budleigh Salterton Pebble Beds took place on a piedmont alluvial fan or alluvial plain (Smith, 1990; Smith and Edwards, 1991). The formation there consists of the deposits of low-sinuosity, northward-flowing, braided channels, with a coarse, pebbly fill.

Suitable British sources of the pebbles of Carboniferous Limestone contained in the formation are to the north and east, and include the southern Mendips and other areas around the Bristol Channel, from Somerset northwards to Dyfed (Riley, 1995). The age (late Famennian to early Viséan) precludes derivation from the younger (Asbian–Brigantian) Westleigh Limestones of mid Devon.

## OTTER SANDSTONE FORMATION

Strata assigned here to the Otter Sandstone Formation occur in three areas of the Minehead Basin: around Alcombe, in the old part of Minehead town, and at the western end of the basin between Periton, Bratton, Wydon Farm and Hindon Farm (Figures 18, 20, 21). Their stratigraphical position is not certain in every case; it is possible, as suggested by Thomas (1940), that some may be marginal facies of the Mercia Mudstone.

At Alcombe, south of Minehead (Figure 21), the attribution of the sandstones to the Otter Sandstone is fairly clear; they rest on inferred Budleigh Salterton Pebble Beds, are succeeded by the Mercia Mudstone, and are lithologically similar to strata in the main Otter Sandstone outcrops south-east of the district (Edmonds and Williams, 1985). West of Minehead, between Bratton and

Hindon Farm, the sandstones apparently underlie the Mercia Mudstone but, where their base is seen, they unconformably overlie Devonian sandstones, without intervening Budleigh Salterton Pebble Beds (Figure 20).

Around Alcombe, the Otter Sandstone rests unconformably on Devonian Hangman Sandstone, but about 200 m west of Staunton Quarry it is overlapped by the Mercia Mudstone; this, in turn, rests on the Hangman Sandstone for about 2 km, as far west as Periton House, near where the Otter Sandstone reappears from beneath the Mercia Mudstone [SS 9510 4567]. The base of the Otter Sandstone is faulted out between Bratton and near Hindon Farm, but in three small, partly faulted outliers near Hindon Farm [SS 9324 4651], west of Wydon Farm [SS 9366 4698], and at Dean's Lane [SS 927 467], the formation apparently rests on Hangman Sandstone (Figure 20).

In the Alcombe area, the Otter Sandstone is estimated to be up to 30 m thick, although only 9 m have been recorded directly (Thomas, 1940, p.21). In 1995 it was best seen in Staunton Quarry, where it consists of yellowish brown, reddish brown and greyish green, locally mottled, moderately cemented, calcareous, silty, fine- to medium-grained sandstone. The cement is predominantly calcite, and weathered surfaces locally show a characteristic honeycomb effect; analysis of two samples showed calcium carbonate contents of 35.4 per cent and 38.3 per cent by weight. The junction with the underlying Budleigh Salterton Pebble Beds is visible in Staunton and Alcombe quarries. At Staunton Quarry [SS 9742 4493], the basal Otter Sandstone consists of pale reddish brown, well-cemented sandstone, with a sharp junction with the underlying Pebble Beds. In the central part of Alcombe Quarry, exposures [SS 9762 4483] show 1.2 m of moderately cemented, mottled pale yellowish brown and reddish brown, fine- to medium-grained, very sparsely pebbly sandstones which may be Otter Sandstone overlying the Budleigh Salterton Pebble Beds.

A thin section (E 70314) of Otter Sandstone from the western end of Staunton Quarry [SS 9723 4492] showed the rock to be very fine grained and calcareous, with some medium and coarse grains, just grain supported in an inequigranular microspar and micritic calcite cement. The detrital grains are mainly quartz, though some potassium feldspar is present, and most have a thin veneer of iron oxide or hydroxide. Another thin section (E 70309), from the north side of Staunton Quarry, showed the rock to be poorly sorted, fine- to very coarse-grained calcareous sandstone, composed of detrital quartz, cherts, cornstone clasts and limestone clasts, in an inequigranular sparry calcite cement. The detrital grains generally have point and simple long contacts, showing little evidence of compaction, and suggesting that the calcite cement is early diagenetic, probably calcrete. Most of the detrital grains have an iron-oxide or hydroxide coating.

The stratigraphical position of sandstones, breccias and conglomerates outcropping in Minehead between Higher Town and The Parade is less clear than those at Alcombe. Thomas (1940, p.35) considered that the Minehead sandstones were 'marginal deposits lying below, and partly equivalent to, the Red Marls [now the Mercia Mud-

stone], formed by the destruction of the Devonian of North Hill, uplifted along the Minehead fault.' In the absence of biostratigraphical evidence, and in view of the wholly fault-bounded nature of the outcrop, the stratigraphical relations of these sandstones and breccias remain uncertain; on the map they are included in the Otter Sandstone. In the few exposures, mainly along Holloway Street (see below), they differ from the predominantly mottled sandstones at Alcombe, and consist mainly of reddish brown, cross-bedded, well-sorted, fine- to medium-grained sandstones with units of fine breccia. A small outcrop of conglomerate [SS 9679 4663] in The Ball was compared by Ussher (BGS MS) with the Alcombe conglomerate (Budleigh Salterton Pebble Beds), but it contains no limestone pebbles and is probably a lenticular bed within a predominantly sandstone unit. It was considered by Thomas (1940, p.18) to be 'no more than a local variant of the breccias and sandstones'. The Holloway Street Borehole [SS 9673 4637] (Figure 22) penetrated 83.7 m of the sequence, but is difficult to interpret because of the existence of conflicting records. Interpretation of the gamma-ray log suggests that the strata consist mainly of sandstone and 'marly' (clayey) sandstone, with units of 'marl' and 'sandy marl' up to 9 m thick which are readily distinguished on the log.

The sedimentary features of the Minehead Otter Sandstone sequence are well displayed in the Holloway Street section (southernmost exposure [SS 9674 4637]), shown graphically in Figure 23. The southern part of the section comprises approximately 9 m of fine- to medium-grained sandstone, dipping north at 12°. Most of the outcrop comprises weakly cemented, well-sorted, reddish brown, fine- to medium-grained sandstone, with moderate to high-angle sets of cross-bedding, the azimuths of which are directed towards 110° The sets are up to 0.6 m thick, typically planar, and with asymptotic bases; rare trough cross-beds up to 0.3 m thick are present. The sandstone is moderately to well sorted, and comprises subangular to subrounded, frosted grains. The foresets comprise alternations of fine- and fine- to medium-grained laminae; rare medium- to coarse-grained laminae up to 8 mm thick occur. Some 'pinstripe' lamination occurs along the lower, asymptotic part of some sets. A few disturbed and upturned foresets occur.

Interbedded with the sets of cross-bedded sandstone are parallel-laminated sandstones and claystones, in successions up to 0.6 m thick. This second type of sandstone occurs in beds up to 0.2 m thick, characterised by the presence of a well-defined, thin, parallel or low-angle ('pinstripe') lamination. The laminae, up to 1 mm thick, comprise alternations of fine- and medium-grained sandstone, in places only one or two grains thick. Some low-angle, cross-bedding foresets are present. In addition, ripple form sets also occur, with no foresets preserved. The form sets are asymmetrical, have a wavelength of 5 to 6 cm, a relief of 8 to 10 mm and are commonly draped by thin claystone laminae. Claystone laminae up to 3 mm thick also occur interbedded with the pinstripe laminated sandstone. They are commonly broken in places by sand-filled desiccation cracks.

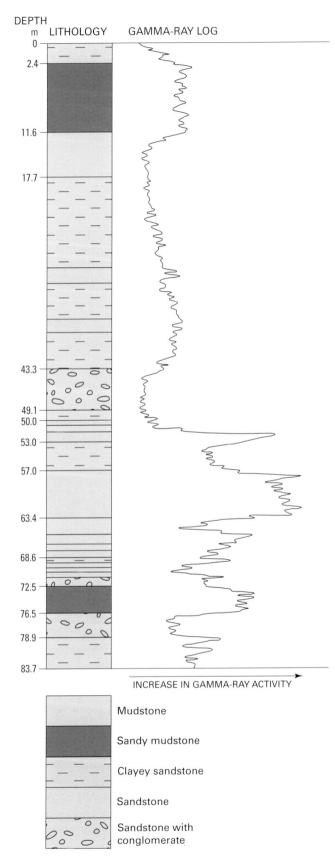

DEPTH
m   LITHOLOGY   GAMMA-RAY LOG

INCREASE IN GAMMA-RAY ACTIVITY

Mudstone

Sandy mudstone

Clayey sandstone

Sandstone

Sandstone with
conglomerate

**Figure 22**   Gamma-ray log and interpretive graphic section of the Otter Sandstone Formation in the Holloway Street Borehole [SS 9673 4637], Minehead.

About 1 m above the base of the section is an erosion surface overlain by 1.85 m of reddish brown, fine- to medium-grained pebbly sandstone. The sandstone is trough cross-bedded, in sets up to 0.5 m thick; the foresets are typically high angle. Angular clasts occur in discrete layers, and vary in size up to medium pebble; the largest clast is $9 \times 5$ cm in size. The clasts are typically composed of fine- and medium-grained sandstone, with some rare vein quartz and intraformational mudstone.

North of a retaining wall, which is about 14 m long, the section continues [SS 9675 4642 to 9676 4645] northwards to a point about 10 m south of the junction of Holloway Street and Clanville Road, and consists mainly of breccias about 3 m thick, with a few impersistent 2 to 3 cm-thick lenses of reddish brown silty sandstone. The breccias are reddish brown, moderately cemented, poorly sorted, clast-supported, and composed of angular, small to medium pebble-sized clasts, the largest $8 \times 4$ cm. The clasts are mainly of Devonian sandstone, dark purple-grey, flat-lying, fissile mudstone, and some vein quartz, in a fine-grained, poorly sorted sandstone matrix. The breccias are trough cross-bedded, with fine-grained sandstone forming the upper parts of sets in places. The sets are up to 0.4 m thick and have foresets directed towards the west. Several small faults affect the section. Owing to the presence of faults in the observed section, and probably in the area of the retaining wall, the relationship of the breccias to the sandstones is not clear.

At the western end of the Minehead Basin (Figures 18, 20) the best exposures (1995) are in Wydon Farm Quarry [SS 9394 4691], which shows about 6 m of weathered sandstone with minor breccia, dipping north at about 20° towards the faulted northern margin of the basin. At the base is about 0.7 m of cross-bedded, fine- to medium-grained sandstone, with rare granules and small pebbles. Cross-bedding sets are up to 0.22 m thick, with foresets directed to the north-east. The sandstone is overlain, above an erosional surface, by 0.23 m of cross-bedded granular to pebbly sandstone. The uppermost unit in the quarry comprises 2.6 m of reddish brown, weakly cemented, moderately sorted, fine- to medium-grained sandstone with a basal breccia bed. Angular clasts up to large pebble size occur, either scattered, in layers, or more rarely forming lenses. The clasts comprise dominantly sandstone and rare vein quartz. A number of small-scale sets of (possibly planar) cross-bedding have low-angle foresets, dipping to the north-east. The basal breccia, 0.38 m thick, is oligomictic, clast-supported, and composed of angular, very large pebble-sized clasts of reddish grey and fine- to medium-grained sandstone in a matrix of dark reddish brown, fine- to medium-grained sandstone. Rare vein quartz clasts also occur.

East of Minehead, boreholes along the south side of the Butlins 'Somerwest World' holiday camp penetrated sandstones and mudstones beneath up to 12.5 m of drift deposits (Figure 24). The stratigraphical relationships of these strata are uncertain. Sandstones in boreholes SS 94 NE/31–34 overlie mudstones which are at least 9.75 m thick in borehole SS 94 NE/34; if the sandstones are interpreted as Otter Sandstone, the mudstones must be interpreted as within that formation. Support for this is

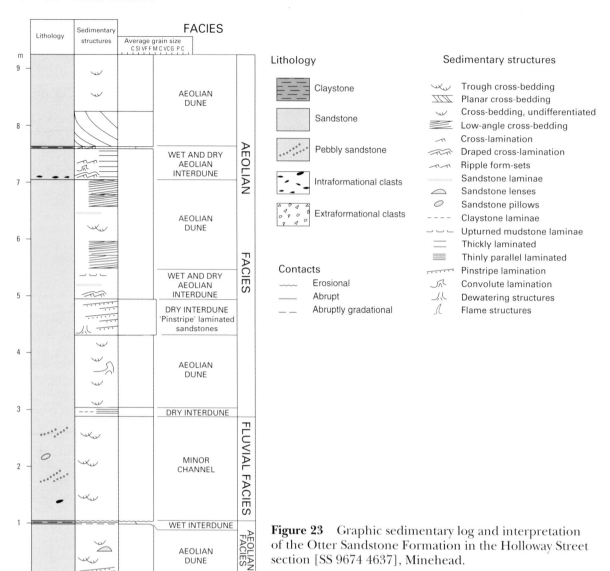

**Figure 23** Graphic sedimentary log and interpretation of the Otter Sandstone Formation in the Holloway Street section [SS 9674 4637], Minehead.

provided by the presence of mudstone units up to 9 m thick within the probable Otter Sandstone of the Holloway Street Borehole (Figure 22). Another possibility is that the sequence is Mercia Mudstone with units of sandstone. In boreholes SS 94 NE/8 and 9, at the south-eastern corner of the site, the beds proved are predominantly mudstone with units of sand and are more readily interpreted as Mercia Mudstone.

The sandstones of the Porlock Basin are included for the purpose of this account in the Mercia Mudstone, and are described under that heading. Sandstones occurring beneath the Mercia Mudstone in the Porlock Basin were termed '"Keuper" Sandy Limestones' by Thomas (1940) and equated with the '"Keuper" Sandstones' of the Minehead area (Otter Sandstone of this account and the 1:50 000 Series map). The resemblance of some of the sandstones west of Minehead (e.g. at Hindon) to those at West Luccombe in the Porlock Basin led Thomas to suggest a correlation, indicating that at this time connection had been established between the Porlock Basin and the

Minehead area. This appeared to be confirmed by the similarity in the heavy mineral suites in both basins. The suite differs from that in the Luccombe Breccia and is characterised by large rounded apatite and dodecahedrally cleaved garnet, together with anatase and possibly monazite (Thomas, 1940).

There is no biostratigraphical evidence for the age of the Otter Sandstone in the district. The nearest dated localities are on the south Devon coast, some 37 km to the south, where the formation is assessed as early Mid-Triassic (Anisian) in age on the basis of vertebrate faunas (Benton et al., 1994).

Further details are given by Edwards (1996).

## Depositional environments

The limited exposures of the Otter Sandstone in the district indicate that deposition took place within fluvial channels, as recorded at Staunton Quarry and Wydon Farm Quarry. The channels were from 2 to 3 m deep and

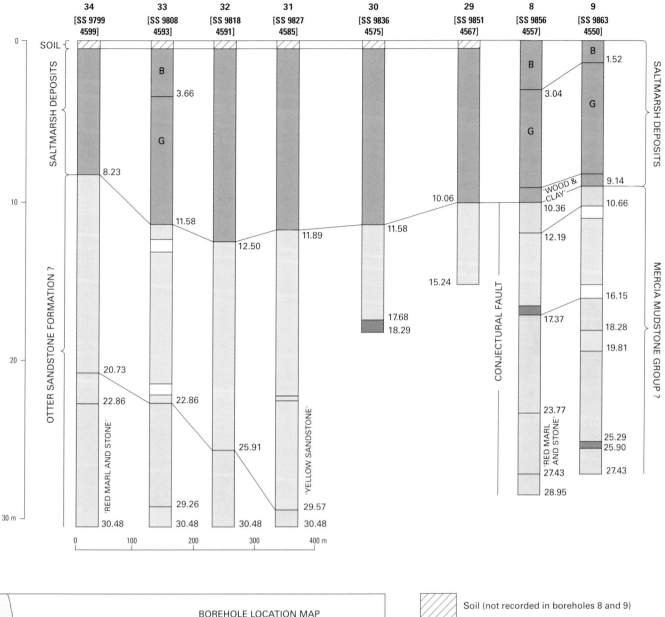

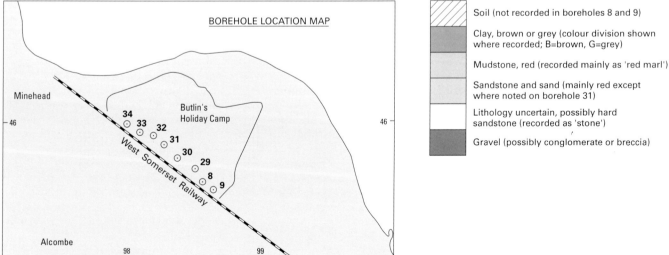

**Figure 24**    Graphic logs of boreholes near Butlin's 'Somerwest World' Holiday Camp, Minehead showing possible equivalents of the Otter Sandstone Formation and the Mercia Mudstone Group overlain by saltmarsh deposits. Borehole numbers refer to BGS archives, in which each number is preceded by SS 94 NE.

had variable, unidirectional flow, generally towards the north. They were filled by small, sinuous-crested dunes and dominantly carried a sandy bedload, although coarser pebbly sediment was also available for transportation. The channel deposits are typically multistorey.

The fluvial deposits are interbedded in the Holloway Street section with aeolian sandstones. Cross-bedded sandstones represent the deposits of small aeolian dunes, up to at least 0.3 m high. This is suggested by the high-angle nature of the foresets and the distinctive foreset stratification styles present. A limited number of measurements indicate that winds blew from west to east. The interbedded pinstripe-laminated sandstones and claystones are interpreted respectively as dry and wet interdune deposits. The pinstripe lamination is formed by the migration of wind ripples; ripple foresets are generally not preserved, although rare wind ripple form sets have been recorded. The wet interdune deposits are represented by deposition of mud; these probably formed in localised pools of water and were temporary features that dried out, resulting in desiccation of the clay laminae. These aeolian deposits probably formed on the alluvial floodplain and may represent reworking of fluvial deposits; reworking is indicated by the presence, in the aeolian sandstones, of grain types similar to those filling the channels, and by the general absence of coarse, millet seed grains.

## MERCIA MUDSTONE GROUP

In the district, the Mercia Mudstone Group ('Keuper Marl' of earlier classifications; Warrington et al., 1980) is not yet fully divided into formations. Unnamed, predominantly reddish brown, dolomitic mudstones and silty mudstones, with some grey-green dolomitic mudstones, sandstones and evaporites, are here referred to as the 'red Mercia Mudstone'. The overlying Blue Anchor Formation, at the top of the group, comprises mainly grey and green mudstones with subordinate reddish brown mudstones, black shales, fine-grained sandstones and evaporites. Its type section lies within the district. The 'Dolomitic Conglomerate', which occurs as a marginal facies in the Mendips area (Green and Welch, 1965; Kellaway and Welch, 1993), is not developed in this district, except possibly for a minor outcrop of breccio-conglomerate near Dunster. However, the possibility that sandstones and breccias in and west of Minehead represent marginal facies of the Mercia Mudstone, as suggested by Thomas (1940), cannot be excluded.

The Mercia Mudstone has two main onshore areas of outcrop within the district: in the Porlock Basin, and between Minehead and Watchet. It crops out offshore, on the sea bed, in a band parallel to and a few kilometres north of the coastline, and in two small inliers at the eastern margin of the sheet (Figure 18), it also has an extensive subcrop offshore beneath younger Triassic and Jurassic rocks. Owing to the small number of sea-bed samples (Figure 18), it is not possible to separate the Mercia Mudstone from the Penarth Group in the offshore area, and the outcrop shown on the map includes rocks of both groups.

The greatest known thickness of the Mercia Mudstone onshore in the district is the 350 m estimated to be present in the Porlock Basin. In the offshore area, the Mercia Mudstone is estimated to be up to 375 m thick. The Blue Anchor Formation is up to about 36.5 m thick.

In the Porlock Basin, the Mercia Mudstone forms the low-lying, largely drift-covered, ground that extends from Porlock Bay eastwards through Holnicote to Tivington. East of Horner, the outcrop lies between the Luccombe Breccia hills to the south and the Penarth Group escarpment to the north (Figure 20); west of Horner, the breccias are absent, and the Mercia Mudstone plain is bordered to north and south by hills of Devonian Hangman Sandstone. At Huntscott, the Mercia Mudstone is faulted down to form part of a more extensive outcrop outside the district, south of Wootton Courtenay.

There are few exposures of the red Mercia Mudstone in the Porlock Basin. About 3 m of typical reddish brown mudstones are exposed in a road cutting [SS 8971 4618] near West Luccombe, and about 10 m of reddish brown, blocky weathering mudstone with scarce centimetre-scale shaly clay beds and a few pale greyish green mudstone beds are exposed in a river-cliff section [SS 8982 4696] about 800 m north of West Luccombe.

Where the Mercia Mudstone overlies the Luccombe Breccia, the nature of the contact is uncertain, owing to poor exposure. The breccias dip northwards at between 30° and 54°, the average of 15 dips being 40°. Dips in the lower part of the overlying mudstones range between 20° and 30°, the average of 10 dips being 25°. This apparent discrepancy suggests the possibility of an angular unconformity between the breccias and the overlying mudstones. However, Thomas (1940, pp.30–31) considered that there was a downwards transition from the basal red mudstones of the Mercia Mudstone, via 'breccia-sandy-limestones', into the Luccombe Breccia.

East of the north–south fault which passes just east of Horner, the Mercia Mudstone overlies the Luccombe Breccia; west of that fault, it rests directly on Devonian sandstones with no intervening breccias (Figure 20). East of Huntscott, the Mercia Mudstone again rests directly on Devonian sandstones east of a north–south fault, the Luccombe Breccia being absent. These variations may reflect erosion after a period of faulting and before deposition of the Mercia Mudstone.

On the north side of the Porlock Basin, between Tivington Heights and Allerford, the Mercia Mudstone is overlain conformably by the Penarth Group; north-west from Allerford, to the coast near Bossington, there is probably a faulted contact between the Mercia Mudstone and Devonian rocks (Figures 18, 20).

Beds of reddish brown sandstone are present in the basal part of the group in the Porlock Basin, particularly in faulted outcrops just east of Horner (Figure 20). The relationship of the sandstones at or close to the base of the Mercia Mudstone in the Porlock Basin to the Otter Sandstone of the Minehead Basin is uncertain in view of the physical separation of the basins and the lack of biostratigraphical evidence.

A thicker sandstone unit, partly exposed at Rydons Quarry [SS 9100 4555] in the Ebshill area, occurs at the base of the group, apparently resting on the Luccombe Breccia near Ebshill Lodge [SS 9108 4546]. The sandstones at Rydons Quarry are poorly sorted, with some well-rounded, frosted grains and rare scattered quartz granules. A few beds have small, pebble-sized mudstone clasts. Each bed has sharp lower and upper boundaries, and is structureless. A thin section (E 70315) showed the sandstone to be fine to very coarse grained, calcareous, and composed of dominant quartz with minor potassium feldspar, grain supported in an inequigranular xenotopic calcite cement. Most of the grains have a thin veneer of iron oxide or hydroxide. The coarse lithic fragments are mainly quartz, with some mudstone clasts and carbonate clasts.

Also present in this area is a distinctive mappable bed of calcareous sandstone which is commonly decalcified at outcrop. A thin section (E 70320) [SS 9045 4544] of the fresh rock shows it to be fine- to medium-grained calcareous sandstone, composed of dominant quartz, with minor potassium feldspar, just grain-supported in an inequigranular xenotopic calcite cement, locally poikilotopic. The cement has inclusions of dolomite rhombs.

Analysis of seven samples of sandstones outcropping at or near the base of the Mercia Mudstone in the Porlock Basin shows calcium carbonate contents of between 22.5 and 53 per cent by weight. Only one analysed sample contained more than 50 per cent calcium carbonate, and thus the description of these strata by Thomas (1940) as 'limestones' (included within his '"Keuper" Sandy Limestones') was inappropriate.

West of the Horner Fault, sandstones at the base of the Mercia Mudstone at West Luccombe have been included in the group, but their stratigraphical position is uncertain, especially since they show some petrological similarities to sandstones in the lower part of the Luccombe Breccia at Huntscott (p.59) (Strong, 1995). However, they were included in the '"Keuper" Sandy Limestones' by Thomas (1940), who distinguished them from the Luccombe Breccia by differences in their heavy mineral content. The sandstones are seen in a quarry [SS 8985 4613] at Burrowhayes, West Luccombe (Figures 20, 25). The exposed section comprises interbedded sheet sandstones up to 0.7 m thick and with sharp bases, and thicker sandstones, up to 2.0 m. Both rest on erosion surfaces and contain trough cross-bedding, cross-lamination and parallel-lamination/plane bedding. Cross-bedding sets are up to 0.3 m thick, with unidirectional azimuths directed generally to the south-west. One example of a dewatering or flame structure is present. The beds appear to become thinner and less pebbly upwards. The sandstones are poorly sorted, and pebble-sized clasts are common. The grains are typically subangular to subrounded, with an abundant haematite grain coating; some coarse grains and granules occur, and rare coarse, frosted, well-rounded aeolian grains are present. Clasts, mainly of limestone, vein quartz and sandstone, are angular to subangular, up to medium pebble size, although averaging granule grade. They are common in the lower parts of sandstone beds. The clasts are generally of low sphericity, with a south-dipping imbrication present in places. The largest clast is up to 0.16 × 0.1 m in size. Silty clay-

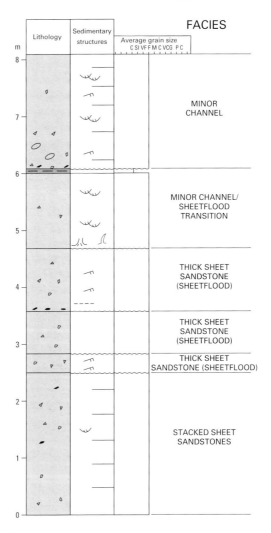

**Figure 25** Graphic sedimentary log and interpretation of basal Mercia Mudstone Group sandstones at Burrowhayes Quarry [SS 8985 4613], West Luccombe. See Figure 23 for key.

stone and siltstone occur in a few beds up to 0.08 m thick.

A thin section (E 70308) shows the rock to be very coarse-grained calcarenite or calcareous sandstone composed of detrital limestone clasts with cherts and quartz grains, grain-supported in a coarse calcite spar matrix. The detrital grains have a pronounced veneer of iron oxide or hydroxide.

In the Minehead Basin, only red Mercia Mudstone is present, as faults have cut out the higher parts of the sequence, and no Blue Anchor Formation has been mapped. The best exposures of the red Mercia Mudstone in the Minehead town area are along Periton Lane [SS 9593 4618 to 9568 4595]. The 350 m-long section is in a sunken lane up to 5 m deep, showing intermittent exposures of mainly reddish brown mudstones, with sporadic thin beds (up to about 0.6 m) of very pale grey calcareous sandstone. The Mercia Mudstone west of the town (between Wydon Farm and Bratton, and near Periton House) rests on Otter Sandstone, and dips northwards towards the boundary fault of the Minehead Basin. Between Periton House and Alcombe, the Mercia Mudstone rests directly and unconformably on the Devonian. The former Alcombe Brickpit (Figure 21) shows [SS 9708 4530] up to 12 m of reddish brown mudstone with a few beds of calcareous sandstone up to 0.25 m thick. A thin section (E 70303) [SS 9719 4525] shows the sandstone to be very fine grained and calcareous, with rare medium and coarse grains of detrital quartz, carbonate clasts (cornstone fragments), and minor feldspar in a micrite/microspar calcite matrix. The grains show point and simple long contacts, indicating minimal compaction, and suggesting that the carbonate matrix is an early near-surface cement (e.g. calcrete).

Near Alcombe, the Otter Sandstone reappears from beneath the Mercia Mudstone. Around Dunster and westwards to Penny Hill, the Mercia Mudstone fills a palaeovalley in which marginal facies of the group are present.

Near Dunster, probable marginal deposits of breccio-conglomerate and sandstones, locally mottled, are present within a predominantly mudstone sequence. The breccio-conglomerates form a ridge extending from near Butter Cross, through Dunster, into Dunster Old Park; they consist of poorly bedded, coarse breccio-conglomerate with subrounded to subangular clasts up to 0.3 m (many coarser than 0.1 m) of purple, quartzitic, fine- to medium-grained Devonian sandstone and some vein quartz in a matrix of reddish brown calcareous sandstone. West of Dunster, the Mercia Mudstone occurs in an elongate hollow between hills of Devonian Hangman Sandstone. There are few exposures, but the indications are that the group is predominantly reddish brown clay and mudstone, and reddish brown sand and sandstone, locally mottled.

Hills such as Conygar Hill represent exhumed features of the pre-Triassic landscape, and probably stood up as hills during deposition of the Mercia Mudstone. The present topography represents partial exhumation of the Triassic topography, a situation recorded also in the adjacent Weston-super-Mare district to the east (Whittaker and Green, 1983)

Between Minehead and the Blue Anchor–Watchet area, the Mercia Mudstone extends beneath the largely drift-covered coastal flats between Alcombe, Marsh Street and Blue Anchor (Figure 18). From Blue Anchor to Watchet (Figures 18, 31) the group is exposed in the cliffs and foreshore, and this area includes the type section of the Blue Anchor Formation (Front cover) (see below). The cliffs [ST 0337 4356 to 0391 4371] immediately east of Blue Anchor expose reddish brown mudstones. Poorly to moderately developed bedding can generally be picked out by alternations of better-bedded, massive mudrock and poorly bedded, more disrupted mudrock. Four main lithofacies are recognised (Jones, 1995):

*Lithofacies 1* is better bedded and generally darker than the other lithofacies, and consists of dark reddish brown to more chocolate-coloured, lithologically homogeneous, silty claystone and siltstone. Beds are generally from 0.1 to 0.3 m thick, and have sharp bases and tops. In places they pass laterally into lenticular units, the upper surfaces of which are locally irregular in form and in places apparently scoured. Low-angle bedding plane shears are present within this facies, and a 'cleavage-like' near-vertical fabric is locally developed. The mudrock is internally massive, with some vugs.

*Lithofacies 2* consists of poorly bedded, vivid brick red mudrock. It is typically finer grained than lithofacies 1, and consists of a mixture of claystone and silty claystone. It is characterised by a rubbly or 'crumb' fabric in which the rock consists of small (less than 1 cm) flattened fragments of mud. Green laminae formed by the reduction of iron are common within the matrix. Desiccation cracks are present in places, and are filled with darker mud, similar to that of lithofacies 1. Small vugs are very common.

*Lithofacies 3* occurs commonly, and is a mudstone breccia resulting from modification of lithofacies 1 and 2. Small angular pebbles of mudstone form the clasts within the breccia. Bedding surfaces reveal the presence of polygonal desiccation cracks with complex, cross-cutting infills. Desiccation resulted in complete brecciation of the sediment.

*Lithofacies 4* comprises brecciated claystone with common remnant evaporite (possibly after gypsum) nodules and wavy, convoluted bedding. Rare anastomosing gypsum veins occur subparallel to bedding. Numerous green laminae are associated with the facies.

Most of the structures and textures seen in the Mercia Mudstone in the Blue Anchor cliffs, such as desiccation cracks, crumb-fabric and breccia, are the result of modification of the original rock fabric by desiccation processes, evaporite growth and pedogenesis. These modifications obscure the original depositional fabrics, rendering interpretation of depositional environments difficult.

The massive nature of *lithofacies 1* makes interpretation difficult. There are no obvious indications of post-depositional modification, but it cannot be ruled out. The lack of lamination argues against an origin by suspension-settling in a subaqueous environment. Formation by deposition of wind-blown silt on a damp substrate is possible, but considered unlikely because the sediment is apparently too well sorted, lithologically homogeneous and lacking in the mixture of lithologies that would characterise such a deposit. It is considered most probable that lithofacies 1 represents the deposits of debris or mud flows; the abundance of mud suggests that the flows were cohesive in nature.

The crumb fabric which characterises *lithofacies 2* is interpreted as pedogenic in origin. The mud fragments are considered to have formed by a combination of desiccation and the formation of peds (stable aggregates of soil material, usually surrounded by a network of irregular planes termed cutans (Retallack, 1988)). It is thought likely that *lithofacies 3* was formed by processes similar to those affecting lithofacies 2, although the processes — particularly desiccation — were more intense and resulted in complete brecciation of the rock. The brecciation was probably the result of extensive wetting and drying and desiccation crack development. The vugs present in most lithofacies are interpreted as hollows formed by the dissolution of evaporite minerals. Sediment may have been supplied by sheetfloods.

The association of *lithofacies 4* with evaporite nodules suggests that some of the brecciation associated with the facies was caused by evaporite growth and dissolution. The formation of evaporites indicates hot arid conditions in which evaporation and precipitation of sulphates occurred.

The red Mercia Mudstone is in 'normal' facies between Blue Anchor and the area of the Watchet Fault [ST 0607 4332], but farther east it locally contains abundant veins and nodules of gypsum. The uppermost 50 m of the red Mercia Mudstone, exposed east of Watchet Harbour, is characterised by interbeds of carbonate within a predominantly red mudstone sequence. The lithology and petrography of these beds have been studied by Talbot et al. (1994) who noted that horizontal bedding is widespread, but difficult to discern, and suggested that primary centimetre- to decimetre-scale horizontal lamination had been disrupted by post-depositional pedogenic processes. The claystones generally contain a mixture of angular to subangular quartz with subordinate feldspar and a range of types of carbonate grains, characteristically 20 to 80 μm dolomite rhombs. Other carbonate grains include limestones, some fossiliferous, and fragments of dense micrite; the coarser clastic grains are set in a matrix of, and commonly impregnated by, red, very fine silt and clay. An indistinct clotted texture in the mudstones suggests that peloids may originally have been common (Talbot et al., 1994). The clay mineral assemblage is dominated by illite, chlorite, mixed-layer chlorite/smectite (corrensite) and smectite; palygorskite/sepiolite is present in small amounts (under 2 per cent of the clay mineral species) at some horizons (Leslie et al., 1993, fig. 6).

## Blue Anchor Formation

The Blue Anchor Formation is the highest formation of the Mercia Mudstone Group and comprises the strata between the predominantly red mudstones that form the bulk of the Mercia Mudstone, and the overlying Westbury Formation of the Penarth Group. The type section is located within the present district at Blue Anchor Cliff [ST 0385 4368] (Warrington and Whittaker, 1984, fig. 2). The term was introduced (Warrington et al., 1980) to include beds formerly comprising the 'Tea Green Marl' and overlying 'Grey Marl' of earlier classifications. It corresponds to Unit F recognised on borehole geophysical logs of the Mercia Mudstone (Lott et al., 1982). The succession consists of predominantly grey and green mudstones, with sulphate evaporites. The thickness at the type section is 36.54 m (Figure 26); in the Porlock Basin, the thickness is estimated at 25 m.

The base of the formation is placed at the top of the highest prominent bed of red mudstone in the underlying red Mercia Mudstone. The upward change in colour from red to grey and green is probably an indication of the higher initial organic content of the Blue Anchor

Formation sediments, which would have the effect of promoting anaerobic diagenetic conditions, as compared to the oxidising conditions in the red sediments (Anderton et al., 1979).

The formation is best displayed in the cliff and foreshore outcrops between Blue Anchor Cliff and Watchet (Figure 26). Eastwards from the type locality, gypsiferous Blue Anchor Formation appears in faulted outcrops in the cliffs and foreshore (Figure 31, pp.84–85). The formation occurs in a fault-bounded syncline [ST 0527 4365] on the foreshore about 450 m north of Warren Farm. It is well exposed, but largely inaccessible, in the cliff section [ST 0600 4331] about 850 m east of Warren Farm and just west of the Watchet Fault; there, the gypsiferous units A, B and C recognised at the type locality were identified and measured using a suspended scale and photographing the section from beach level (Warrington and Whittaker, 1984).

On the foreshore about 1 km west of the northern breakwater of the harbour at Watchet, the formation is exposed in the core of an east–west-trending pericline [ST 063 437] (Figure 31, pp.84–85). Further faulted outcrops occur on the foreshore north of Watchet Harbour [ST 073 437], and a more continuous outcrop extends from there eastwards into Helwell Bay, just east of the district boundary. A measured section [ST 0770 4355], some 500 m east of Watchet Harbour and 100 m east of the district boundary, is given in Whittaker and Green (1983, pp.49–50; fig. 10), and is reproduced graphically in Figure 26 for comparison with that at Blue Anchor.

In the Porlock Basin, the outcrop has been mapped for about 3.5 km along the northern side of the Mercia Mudstone outcrop from near Venniford Cross to Allerford (Figure 20); it is concealed beneath Quaternary deposits around Holnicote and Brandish Street, and may continue towards Bossington and Porlock Bay, also beneath superficial deposits. Measured dips range between 10° to the north-north-east near Brandish Street, and 18° north near Buddle Hill, Holnicote.

An incomplete sequence of the upper 13.41 m (corrected for dip) of the Blue Anchor Formation was penetrated in the Selworthy No. 2 Borehole [SS 9244 4618] between depths of 37.62 and 51.44 m (Figure 26). In an earlier account (Institute of Geological Sciences, 1974), the upper boundary of the formation was placed at a depth of 35.61 m, but this has been revised downwards to 37.62 m because the beds between the two depths are lithologically closer to Westbury Formation lithologies, and as a result of comparison of the palynomorph succession recorded from the borehole (Figure 28) with that from the Blue Anchor Formation and Westbury Formation at St Audrie's Bay (Warrington, 1974, 1981, 1983; Warrington and Whittaker, 1984). In the borehole the formation overlies a fault zone (51.44 m and 52.02–52.22 m) which contains reddish brown mudstones and fractured Devonian Hangman Sandstone. The formation consists mainly of siltstones, and grey to grey-green claystones, commonly with silt laminae; veins of anhydrite are present in the lower 7.6 m. A graphic log of the borehole is given in Figure 26.

Mayall (1981) divided the Blue Anchor Formation of west Somerset into the Rydon Member and overlying Willi-

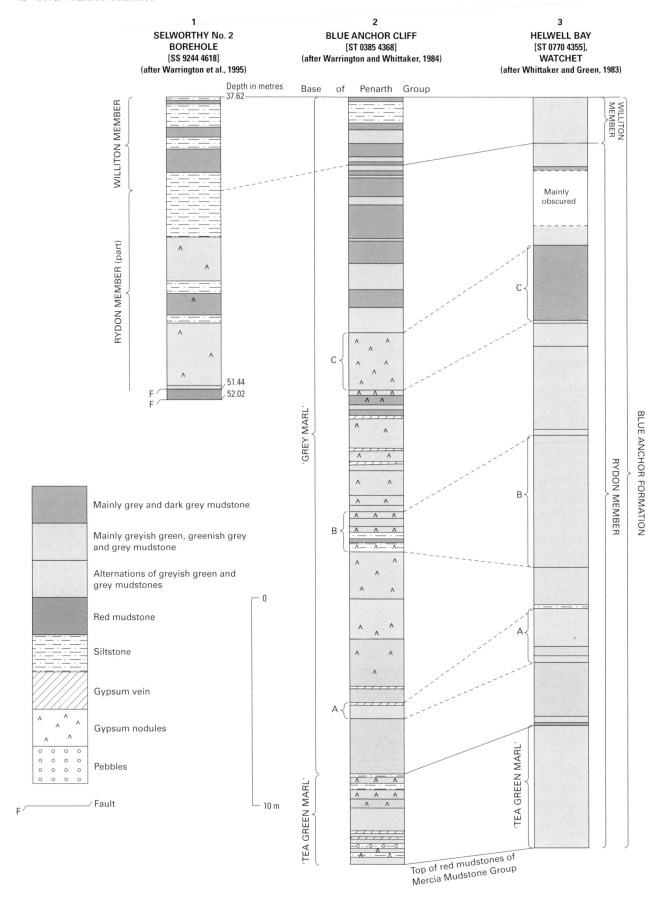

**1**
SELWORTHY No. 2
BOREHOLE
[SS 9244 4618]
(after Warrington et al., 1995)

**2**
BLUE ANCHOR CLIFF
[ST 0385 4368]
(after Warrington and Whittaker, 1984)

**3**
HELWELL BAY
[ST 0770 4355],
WATCHET
(after Whittaker and Green, 1983)

Depth in metres
37.62

Base   of   Penarth   Group

WILLITON MEMBER

RYDON MEMBER (part)

51.44
52.02
F
F

'GREY MARL'

'TEA GREEN MARL'

C

B

A

WILLITON
MEMBER

Mainly
obscured

C

B

A

RYDON MEMBER

BLUE ANCHOR FORMATION

'TEA GREEN MARL'

Top of red mudstones of
Mercia Mudstone Group

Mainly grey and dark grey mudstone

Mainly greyish green, greenish grey
and grey mudstone

Alternations of greyish green and
grey mudstones

Red mudstone

Siltstone

Gypsum vein

Gypsum nodules

Pebbles

0

10 m

F                          Fault

ton Member. The **Rydon Member** constitutes the bulk of the formation and is about 29 m thick at Blue Anchor. It is named after a village in north Somerset near St Audrie's Bay, and the type area was designated by Mayall as the same as that for the Blue Anchor Formation (Warrington et al., 1980). Detailed measurements at the subsequently designated type section of the formation (Warrington and Whittaker, 1984) show that the Rydon Member consists of alternating grey mudstones and greenish grey silty mudstones, locally dolomitic, and with dolomites present especially towards the top of the member. Also present are laminated beds with mudcracks, scarce pseudomorphs after halite, locally burrowed mudstones, and small-scale collapse structures caused by the solution of evaporites. A distinctive feature is the presence of nodular masses and veins of colourless, white and pink gypsum, which occur throughout the member, but are abundant in beds to within about 7.7 m of its top. Mayall (1981, p.377) noted that the gypsum nodules are commonly partly silicified at Blue Anchor. Units especially rich in nodular gypsum were designated B and C by Warrington and Whittaker (1984) (Figure 26) and considered to correspond approximately to beds 12 and 13 respectively of Richardson (1911). Another laterally extensive and correlatable unit, A in Figure 26, consists of 0.05 m of gypsum on 0.69 m of green blocky mudstone with an 0.08 m wispy-bedded unit 0.23 m below the top. Unit A is present in the section 500 m east of Watchet Harbour, and also at St Audrie's Bay [ST 1053 4314], east of the district, where a thin bed of red claystone associated with gypsum overlies a hard grey siltstone bed, commonly with cavities (Whittaker and Green, 1983, pp.50–52). Units A, B and C can also be identified in cliff sections [ST 0600 4331] about 850 m east of Warren Farm; the bases of units B and C are about 22.25 and 10.68 m respectively, below the top of the Blue Anchor Formation, unit B being 1.83 m and Unit C about 2.74 m thick. In this section, Unit A occurs about 28.35 m below the top of the formation. The top of the member is an irregular erosion surface penetrated by *Diplocraterion* burrows which commonly disturb the surrounding sediment.

The **Williton Member,** at the top of the Blue Anchor Formation, ranges in thickness from about 2.5 to about 4 m in the district; there is a westerly thickening between Watchet and Selworthy (Figure 26). The member was introduced by Mayall (1981), who designated St Audries's Bay [ST 103 431] as the type section; it is named after a village [ST 080 410] immediately south of the district. The member overlies the erosion surface at the top of the Rydon Member. A thin intraformational conglomerate is present locally at the base of the member which consists of grey shales with units of flaser- and lenticular-bedded fine-grained sand and silt. Silt-streak beds with desiccation cracks were noted by Mayall (1981, p.379) at

**Figure 26** Generalised graphic sections of the Blue Anchor Formation.

See text, this page, for explanation of units A, B and C in columns 2 and 3.

the top of the member at Blue Anchor and Watchet, and synsedimentary microfaults were also recorded at Blue Anchor.

In Selworthy No. 2 Borehole, the uppermost 4 m of the Blue Anchor Formation has microfloral and microfaunal characters indicating deposition in marine environments, and comparable to those of the Williton Member farther east. This member therefore appears to persist westwards into the Porlock Basin where, at 4 m, it is slightly thicker than at Blue Anchor (3 m), suggesting a continuation of the gradual westerly thickening noted by Mayall (1981).

The clay mineralogy of the Blue Anchor Formation was investigated by Mayall (1979, 1981) who found it to consist of illite and chlorite believed to represent detrital minerals derived from a surrounding landmass under increasingly humid conditions (Mayall, 1979); additionally, corrensite (a mixed-layer chlorite-smectite) was present in the Rydon Member but not in the Williton Member.

## Biostratigraphy

### Mercia Mudstone beneath the Blue Anchor Formation

Two samples, of a total of 14 from the red Mercia Mudstone of the Porlock and Minehead basins, yielded miospores; both were from the Porlock Basin succession. The stratigraphically highest sample (MPA 41672) was collected from small exposures [SS 9233 4586] of very pale grey, hard, shaly claystones with scarce red bands, a few metres below the base of the Blue Anchor Formation near Brakeley Wood, Selworthy (Figure 20). The miospores are yellow and generally fairly well preserved; the following were recognised: *Leptolepidites argenteaeformis*, *?Tsugaepollenites? pseudomassulae*, *Chasmatosporites magnolioides*, *Alisporites* sp., *Ovalipollis pseudoalatus*, *Quadraeculina anellaeformis*, *Vesicaspora fuscus*, *Classopollis torosus*, *Geopollis zwolinskae*, *Gliscopollis meyeriana*, *Granuloperculatipollis rudis* and *Rhaetipollis germanicus*. The presence of *Quadraeculina anellaeformis* and a possible *Tsugaepollenites? pseudomassulae* in the association is indicative of a Rhaetian (late Late Triassic) age.

The second productive sample (MPA 41673), estimated to be 75 m above the base of the group, was collected from siltstone in an exposure [SS 9301 4492] near Troyte's Farm, Tivington (Figure 20). The preparation proved virtually devoid of organic matter, but a few fragments of brown structureless detrital organic matter and two yellow-brown circumpolles were recovered; one of the circumpolles was identifiable as *Gliscopollis meyeriana*, which is indicative of a Carnian to Rhaetian (Late Triassic) age.

### Blue Anchor Formation

RYDON MEMBER    The trace fossil *Arenicolites* is present in the Rydon Member, and *Diplocraterion* burrows penetrate the upper surface of the unit and disturb the highest bed (Mayall, 1981). No marine bivalves have been identified. Miospores have been recorded from the member on the west Somerset coast at St Audrie's Bay, just east of the district (Warrington and Whittaker, 1984, fig. 3), and from the Selworthy No. 2 Borehole (Figure 28). The associations are dominated by pollen of the circumpolles group (*Classopollis*, *Geopollis*, *Granuloperculatipollis* and *Gliscopollis*) and *Ovalipollis pseudoalatus*, but include small numbers of bisaccate and monosulcate pollen and trilete spores. The diversity of the assemblages increases upwards through the member. In the Selworthy No. 2 Borehole, acritarchs (*Micrhystridium*) appear about 4 m below the top of the member.

WILLITON MEMBER    Bivalves have been recorded from the Williton Member by earlier authors, including Boyd-Dawkins (1864) and Richardson (1911). Mayall (1981) noted that shell beds are common in the member, although the bivalves are generally poorly preserved. He listed *Chlamys* sp., *Eotrapezium* sp., '*Gervillia*' *praecursor*, *Modiolus* sp. and *Protocardia* sp. In contrast to the Rydon Member, trace fossils are well preserved along bedding planes at Blue Anchor; ichnogenera present include *Arenicolites*, *Diplocraterion*, *Muensteria*, *Planolites*, *Rhizocoralium* and *Siphonites* (Mayall, 1981).

Palynomorphs recorded from the Williton Member are mostly miospores (Warrington and Whittaker, 1984; Warrington et al., 1995) which comprise assemblages comparable in composition with those from the Rydon Member. Also present are sporadic organic-walled microplankton, including acritarchs and the dinoflagellate cyst *Rhaetogonyaulax rhaetica*, which indicates the Rr Biozone of Woollam and Riding (1983). These records of *R. rhaetica* herald the presence of associations dominated by that dinoflagellate cyst in the overlying Westbury Formation of the Penarth Group.

Two localities in the Blue Anchor Formation of the Porlock Basin yielded palynomorphs (Warrington, 1994). A sample (MPA 41670) from a pit [SS 9055 4666] near Brandish Street (Figure 20) yielded a few yellow, poorly preserved miospores; *Ovalipollis pseudoalatus* and indeterminate non-taeniate bisaccate pollen were recognised. The presence of *O. pseudoalatus* is indicative of a Ladinian (late Mid Triassic) to Rhaetian (late Late Triassic) age. A sample (MPA 41671) from exposures [SS 9208 4598] south of Great Wood, Selworthy (Figure 20), yielded a moderately rich organic residue comprising miospores and detrital organic matter, in similar abundances. The latter comprised yellow and brown, structureless material, and brown and black, structured phytoclasts, including woody fragments; fragments of the colonial alga *Botryococcus* were also present. The miospores, though relatively abundant, are poorly preserved; most specimens are yellow. The following were recognised: *Porcellispora longdonensis*, *Ovalipollis pseudoalatus*, *Classopollis torosus*, *Gliscopollis meyeriana* and *Granuloperculatipollis rudis*. This association is indicative of a Norian to Rhaetian (mid- to late Late Triassic) age.

## Depositional environments

The views cited below on the environment of deposition of the Mercia Mudstone are based on studies of the upper 80 m or so of the group exposed in the west Somerset sections, and are not necessarily applicable to the substantial thickness of lower unexposed beds. Talbot et al.

(1994) summarised the main depositional models currently available for the red Mercia Mudstone seen at outcrop in west Somerset; these include:

i)   Deposition by subaqueous accumulation in lacustrine basins.
ii)  Accumulation of wind-blown sediment on a subaerial mudflat/playa.
iii) Deposition on mud-dominated, low-angle alluvial fans.

Talbot et al. (1994) concluded that deposition of the part of the sequence studied took place in a semi-arid to arid, low-relief, continental basin to which sediment was supplied by a combination of sheetfloods, as wind-blown mud, and by rivers carrying peloidal mud. The results of this survey are in broad agreement with these conclusions, though subaerial deposition is considered less important than subaqueous deposition from sheetfloods. In addition, debris flow processes are also suggested as an important contributor to sedimentation. The depositional setting evidently varied from subaqueous lacustrine to subaerial mudflat/playa.

Some authors (see Taylor, 1983) have considered that marine waters provided the source of the Mg in mudstones of the group. Wright and Sandler (1994) believed that there was no need to invoke a marine origin, and that the Mg-clays could be explained by a model in which changes in the chemistry of shallow groundwater resulted in a range of early diagenetic products, including groundwater dolocretes and Mg clays.

Leslie et al. (1993) noted that $\delta^{18}O$ isotopic variations in the upper 80 m or so of the red Mercia Mudstone Group at St Audrie's Bay suggested a predominantly continental source of waters in the Somerset Basin.

Talbot et al. (1994) emphasised the role of post-depositional, pedogenic processes in the formation of many of the textures present in the Mercia Mudstone. Typical features of soils, such as peds, cutans and argillans, were recognised, together with vertisol features. The sediment surface probably became intermittently evaporitic, and the presence of vugs indicates that some dissolution of evaporite minerals has occurred.

The deposits of sheetfloods and minor channels are also present within the Mercia Mudstone of the district. The sheetflood deposits are dominantly sandy, although channel fills vary from sandy to pebbly. It is likely that these represent deposition on marginal alluvial fans which, during periods of high rainfall, prograded farther into the basin; the exact relationship of these coarser facies within the mudrocks of the Mercia Mudstone cannot, however, be established, owing to outcrop limitations.

The Blue Anchor Formation contains sulphate nodules associated with carbonaceous (possibly algal mat) mudstones which have been taken to indicate supratidal deposition in environments comparable to those on modern marine sabkhas (e.g. Sellwood et al., 1970). The supratidal deposits alternate with low-energy intertidal deposits (e.g. laminated dolomitised siltstones), and these beds may contain marine fossils (Whittaker and Green, 1983; Warrington and Whittaker, 1984; Warrington and Ivimey-Cook, 1995). The sedimentology, fauna and trace fossils indicate that the topmost part of the formation (Williton Member) was deposited in a shallow marine environment (Mayall, 1981). Thomas et al. (1993) considered that the organic geochemistry of the highest 9 m of the Rydon Member reflects an upward passage from supratidal to marginal marine conditions.

## Coastal exposures of the Mercia Mudstone

The best exposures of the red Mercia Mudstone in the district are those in the cliffs and on the foreshore platform between Blue Anchor and Watchet. The distribution of the extensive outcrops in the foreshore zone, with fault and dip data, are shown in Figure 31, pp.84–85. The descriptions below refer mainly to the more accessible cliff sections.

The 0.5 km-long stretch of coastline east of the Blue Anchor Hotel [ST 0332 4355 to 0380 4366] consists of cliffs up to about 25 m high in reddish brown mudstones with sporadic thin (less than ,0.1 m) beds of pale greyish green mudstone, mainly dipping south or south-east and with a few small faults. The lithofacies in this section are described above (p.70). Towards the eastern end of the cliff is a large extensional fault [ST 0380 4366] which brings grey Blue Anchor Formation on the east down against red Mercia Mudstone on the west; owing to the colour contrast this is one of the most striking features in this section of the coast (Front cover).

At Blue Anchor Cliff [ST 0385 4368], the full thickness (c. 36.5 m) of the Blue Anchor Formation is exposed; Penarth Group strata and the basal part of the Blue Lias are present at the top of the cliff (Front cover). This locality is the type section of the Blue Anchor Formation (Warrington and Whittaker, 1984). The sequence consists predominantly of grey and green mudstones, with sulphate evaporites. A graphic log is given in Figure 26, based on the detailed measured section in Warrington and Whittaker (1984). About 80 m north-east of the type section, the cliff [ST 0392 4371] shows a shallow anticline trending approximately east–west in the hanging wall of an extensional listric fault; well-developed conjugate gypsum veins are a conspicuous feature (Bradshaw and Hamilton, 1967) (Plate 22).

Farther east, red Mercia Mudstone is exposed in a small headland [ST 0479 4366] where faulted

reddish brown claystones contain several beds, up to about 0.2 m thick, of pale grey-green calcareous siltstone. Several extensional faults with throws of 0.2 to 0.5 m are present on the north side.

In the headland [ST 0552 4338] about 450 m east-north-east of Warren Farm, reddish brown and grey to pale greyish green mudstones, dipping north at 26°, are faulted against *angulata* Zone Lias Group beds which form the cliff to the south. The cliff [ST 0600 4331] on the west side of the Watchet Fault shows interbedded reddish brown and green mudstones overlain in the top third of the cliff by Blue Anchor Formation. Between the Watchet Fault and West Street Beach, Watchet, the lower cliff is formed of gypsiferous, reddish brown mudstones, faulted against the Lias Group in the upper cliff. About 100 m east of the Watchet Fault, gypsiferous, reddish brown mudstones with some thin green beds [ST 0613 4337] on the north, dip south at 26° and are faulted against gypsiferous Blue Anchor Formation on the south. Eastwards [ST 0635 4342], locally faulted, reddish brown, gypsiferous mudstones form the lower cliff and are in faulted contact with Lias Group strata in the upper cliff. Farther east [ST 0660 4348], the lower 6 m of the cliff consist of gypsiferous, reddish brown mudstones with thin veins and nodules of gypsum subparallel to bedding, and some near-vertical veins; the upper 6 to 10 m of the cliff is predominantly reddish brown mudstone. Farther along [ST 0670 4351], gypsiferous, reddish brown mudstones with a few thin greyish green bands are present in the lower part of the cliff. Small faults show the growth of gypsum along the fault planes and this is also apparent along bedding surfaces (Davison, 1994) (pp.98–99).

The first promontory [ST 0681 4353] west of the slipway in West Street Beach, Watchet, shows, in the lower 8 to 10 m of the cliff, reddish brown mudstones with beds of greyish green calcareous mudstone up to about 0.2 m thick. The upper part of the cliff shows about 5 m of pre-

**Plate 22** Cliff [ST 0392 4371] about 80 m north-east of Blue Anchor Cliff showing shallow anticline in Blue Anchor Formation with conjugate veins of gypsum. (July 1995) (GS502).

Since Plate 22 was photographed, substantial cliff falls have taken place in the vicinity of Blue Anchor Cliff (winter 1995) (see Plate 2).

dominantly grey to greenish grey mudstone with only minor beds of reddish brown mudstone. East of Watchet Harbour, exposures [ST 0730 4353] show up to about 50 m of reddish brown mudstones with thin interbeds of pale carbonate (Talbot et al., 1994, fig. 10 a).

## PENARTH GROUP

Rocks assigned to the Penarth Group (the 'Rhaetic' of earlier classifications; Warrington et al., 1980) are seen sporadically in the cliffs and foreshore between Blue Anchor and Watchet (Figure 31, pp.84–85). They also outcrop along the north side of the Porlock Basin (Figure 20), and were proved in the Selworthy No. 2 Borehole [SS 9244 4618] (Figure 27). The group is present widely offshore beneath the Bristol Channel and reappears to the north, in South Glamorgan. A few cores of Penarth Group lithologies have been recovered from the sea bed, for example, north-west of Porlock (samples 190, 191 and 195, Figure 18), but the group cannot be mapped separately from the Mercia Mudstone in the offshore area, because of the scarcity of sample points, and its limits there are not traceable using seismic data.

In the Porlock Basin, the Penarth Group outcrop lies immediately north of that of the Blue Anchor Formation, and locally gives rise to a small, partly wooded scarp between Brandish Street and Tivington (Figure 20). There are no exposures, but the full thickness of the group was penetrated in the Selworthy No. 2 Borehole between depths of 24.77 and 37.62 m; the latter figure for the base is revised from that given in an earlier account (Institute of Geological Sciences, 1974). The thickness of the group, corrected for a dip of 15°, is 12.46 m, a figure which is about 4.2 m less than that at St Audrie's Bay, about 18 km farther east (Whittaker and Green, 1983). The Penarth Group sequence proved in the borehole is shown graphically in Figure 27.

The Penarth Group is sporadically exposed on the foreshore and in the cliffs between Blue Anchor and Watchet. On the foreshore platform north of Blue Anchor Cliff, the Penarth Group forms a northward-dipping sequence conformable with (but locally faulted against) the Blue Anchor Formation below and the Lias Group above; a prominent, approximately east–west, fault brings the Penarth and Lias group sequences against red Mercia Mudstone (Figure 31, pp.84–85). The group is also exposed, but inaccessible, at the top of Blue Anchor Cliff [ST 0385 4368]. Richardson (1911, pp.15–18) measured a section in the Penarth Group across the foreshore in the vicinity of Blue Anchor Cliff, and this remains the most detailed record of the Penarth Group of the district. Richardson's account greatly amplified those by Boyd-Dawkins (1864), Bristow and Etheridge (1873), and the summary by Woodward (1893). The section is shown graphically in Figure 27. The names given to the fossils by Richardson have been revised on the figure; the bed names have not been changed.

The Penarth Group is exposed on the foreshore platform in Warren Bay on the north limb of a faulted anticline (Figure 31, pp.84–85). The section given in Whittaker

**Figure 27** Generalised graphic sections of the Penarth Group.

Column 3: Helwell Bay [ST 0765 4362] (for Westbury Formation); Doniford Bay [ST 0827 4345] (for Lilstock Formation) (after Whittaker and Green, 1983).

The bed numbers and names in columns 2 and 3 are those of Richardson (1911). The following faunas, for which the nomenclature has been revised, were listed by Richardson (1911) from the beds indicated in the Blue Anchor section (column 2):

**'Langport Beds'**
1–3  *Liostrea hisingeri, Atreta intusstriata*
5    *Atreta intusstriata, Protocardia rhaetica*

**'Cotham Beds'**
2    Ostracods at the top

**'Westbury Beds'**
5a (1): *Atreta intusstriata, Chlamys valoniensis, Rhaetavicula contorta*
5a (2): *Atreta intusstriata, Chlamys valoniensis*
5a (3): *Chlamys valoniensis, R. contorta, Placunopsis alpina*
5b ('*Cardium-cloacinum* Bed'): *Placunopsis alpina, Protocardia rhaetica, R. contorta, Tutcheria cloacina*
6: *Atreta intusstriata, Chlamys valoniensis, Modiolus, Placunopsis alpina, 'Pleurophorus' elongatus, Protocardia rhaetica, R. contorta*
7: *Acteonina fusiformis, Acteonina ovalis, Acteonina oviformis; Chlamys valoniensis, Dacryomya titei, 'Pleurophorus' elongatus*
9: fish scales
13 ('*Pleurophorus* Bed'): '*Chemnitzia*' sp.; *Eotrapezium concentricum, 'Gervillia' praecursor, 'Pleurophorus' elongatus; Nemacanthus* sp., coprolites
15 ('The Bone Bed'): *Eotrapezium* sp.; *Ceratodus latissimus, Hybodus minor, Gyrolepis alberti, Lissodus minimus*
17: *Eotrapezium* sp.
21: *Eotrapezium* sp., *Pteromya crowcombeia*
23: *Protocardia* sp.
25: '*Chemnitzia*' sp., '*Natica*' *oppelii; Eotrapezium* sp., '*Pleurophorus*' *elongatus, R. contorta*
27 ('The Clough'): *Dacryomya titei* (at top); *Gyrolepis alberti, H. minor, Lissodus minimus*, reptile bone
29: fish fragments
32: *Eotrapezium* sp., '*Pleurophorus*' *elongatus, R. contorta*
33: *R. contorta*; fish remains including *Gyrolepis alberti, Polyacrodus, H. minor, Lepidotes* sp., *Lissodus minimus*

and Green (1983), and shown graphically in Figure 27, is from a locality [ST 0765 4362] on the foreshore just east of Watchet Harbour, only 50 m east of the district boundary.

The sequence and nature of the beds is similar to that described for the adjacent Weston-super-Mare district (Whittaker and Green, 1983) where the Penarth Group is divisible into three contrasting lithological units. In upwards sequence, the group comprises the Westbury Formation, of dark grey shales and thin limestones, overlain by the Lilstock Formation. The latter comprises a lower (Cotham) member, of greenish grey calcareous shales and thin buff limestones, and an upper (Langport) member, of hard, grey-brown limestones with thin marly interbeds. The Lilstock Formation is, in turn, overlain by the Lias Group; most of the Watchet Beds of Richardson (1911) are now included in the basal part of the Lias Group (Whittaker, 1978 b; Cope et al., 1980).

Mayall (1981) recognised that the base of the Westbury Formation rests disconformably upon the top of the

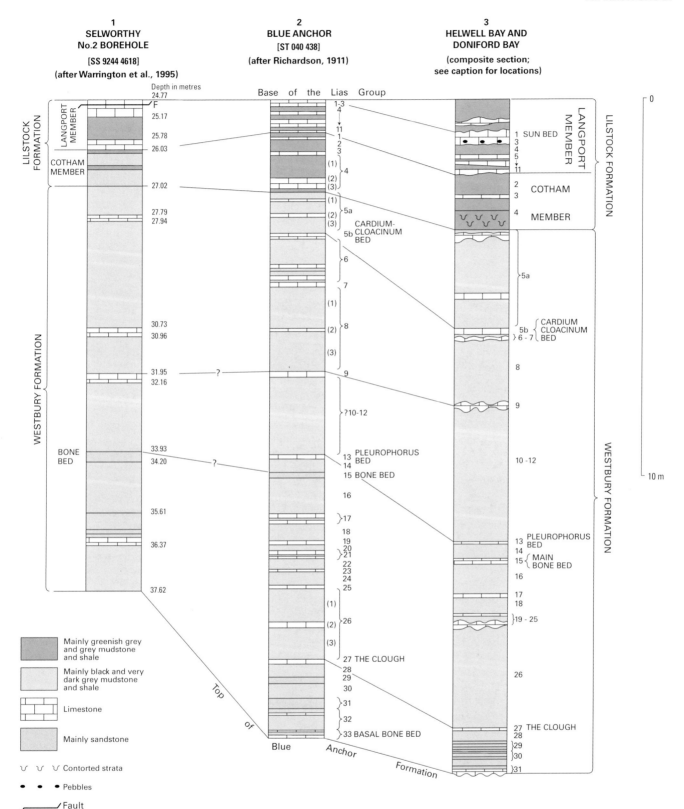

**1**
**SELWORTHY**
**No.2 BOREHOLE**
**[SS 9244 4618]**
**(after Warrington et al., 1995)**

**2**
**BLUE ANCHOR**
**[ST 040 438]**
**(after Richardson, 1911)**

**3**
**HELWELL BAY AND**
**DONIFORD BAY**
**(composite section;**
**see caption for locations)**

Depth in metres

Base of the Lias Group

LILSTOCK FORMATION

LANGPORT MEMBER

COTHAM MEMBER

WESTBURY FORMATION

BONE BED

24.77
25.17
25.78
26.03
27.02
27.79
27.94
30.73
30.96
31.95
32.16
33.93
34.20
35.61
36.37
37.62

Top of Blue Anchor Formation

1-3
4
11
1
2
3
4
5a
5b CARDIUM-CLOACINUM BED
6
7
8
9
?10-12
13 PLEUROPHORUS BED
14
15 BONE BED
16
17
18
19
20
21
22
23
24
25
26
27 THE CLOUGH
28
29
30
31
32
33 BASAL BONE BED

(1)
(2)
(3)
(1)
(2)
(3)
(1)
(2)
(3)
(1)
(2)
(3)

LILSTOCK FORMATION

LANGPORT MEMBER

1 SUN BED
3
4
5
11

COTHAM
2
3
MEMBER
4

5a

5b CARDIUM
6-7 CLOACINUM BED
8
9
10-12

WESTBURY FORMATION

13 PLEUROPHORUS BED
14
15 MAIN BONE BED
16
17
18
19-25
26
27 THE CLOUGH
28
29
30
31

0

10 m

Mainly greenish grey and grey mudstone and shale

Mainly black and very dark grey mudstone and shale

Limestone

Mainly sandstone

ᕼ ᕼ ᕼ Contorted strata

● ● ● Pebbles

⌐___/ Fault

Williton Member of the Blue Anchor Formation along the Somerset coast; the disconformity is marked by an horizon with *Diplocraterion* burrows.

## Westbury Formation

The Westbury Formation sequence recorded by Richardson (1911) in the Blue Anchor Cliff area (Figure 27) comprises dark grey, shaly mudstones with bivalve-rich levels, and also some thin, grey, arenaceous, argillaceous and micritic limestones which are commonly lenticular and associated with fibrous, calcitic 'beef' deposits. Thin lenticles of sandstone, commonly pyritic and with variable amounts of comminuted bone, scales and teeth, also occur at several levels. Richardson (1911) attributed 13 feet 1 inch (4 m) of dark grey shaly marls to the 'Sully Beds', which he included in the 'Lower Rhaetic'; these beds are now included in the Blue Anchor Formation. Above, Richardson recorded 46 feet 3½ inches (14.09 m) of 'Westbury Beds' which he divided using the bed numbering system that he had used for correlation farther east in Somerset and in Glamorgan, where some of these beds appear to be laterally persistent.

In the Selworthy No. 2 Borehole, the Westbury Formation (27.02 to 37.62 m depth) has a corrected thickness of 10.28 m and consists of medium to dark grey mudstones and silty mudstones, with several thin and two thicker limestones (Figure 27). The thicker limestones occur between 30.73 to 30.96 m and between 31.95 to 32.16 m. Also present in the lower part of the formation are some very silty, pyritic and sandy limestones which include abundant fish fragments in silty sand ('Bone Bed') at 33.83 m.

The beds are quite fossiliferous and yielded the gastropod '*Natica*' *oppelii*, and a diverse bivalve fauna including *Atreta intusstriata*, *Cardinia* sp., *Chlamys valoniensis*, *Eotrapezium concentricum*, *Lyriomyophoria postera*, '*Modiolus*' *sodburiensis*, *Protocardia rhaetica* and *Rhaetavicula contorta*. Fish remains include *Lissodus* sp. and indeterminate material. Rich assemblages of palynomorphs, comprising miospores and organic-walled microplankton, were recovered from the borehole samples; their stratigraphical distribution and relative abundances are shown in Figure 28. Circumpolles pollen, principally *Classopollis torosus*, are commonly a major constituent of the miospore associations, but *Ovalipollis pseudoalatus*, *Rhaetipollis germanicus* and *Ricciisporites tuberculatus* occur consistently and in relatively large numbers. *Granuloperculatipollis rudis*, which is relatively common in the Blue Anchor Formation, declines in abundance within the Westbury Formation, and has not been observed at higher levels in the borehole; the distribution of *Geopollis zwolinskae* is similar, with only a doubtful record in beds above the Westbury Formation. The dinoflagellate cyst *Rhaetogonyaulax rhaetica* is present in relatively large numbers in assemblages from beds close above the base of the formation, and dominates some from the upper part of the formation (Figure 28).

The biostratigraphy of the Westbury Formation in west Somerset was reviewed by Warrington and Ivimey-Cook (1995). They noted that the formation yields a macrofauna dominated by bivalves, including *Chlamys valoniensis*, *Eotrapezium concentricum*, *Modiolus* sp., *Protocardia rhaetica*, *Rhaetavicula contorta*, *Placunopsis alpina*, '*Pleurophorus*' *elongatus* and *Tutcheria cloacina* (Figure 30). Other constituents of the fauna are the ophiuroid *Aplocoma* and, mainly from bone-rich beds low in the sequence, the remains of fish (*Birgeria*, *Ceratodus*, *Dalatias*, *Gyrolepis*, *Hybodus* and *Lissodus*) and marine reptiles. Microfloras consist of rich and diverse spore and pollen associations, with organic-walled microplankton, predominantly the dinoflagellate cyst *Rhaetogonyaulax rhaetica* which indicates the Rr Biozone of Woollam and Riding (1983), and occur throughout the formation (Warrington, 1974, 1981, 1983, 1985). In localities east of the district, tasmanitid algae occur sporadically (Warrington, 1981), and coccoliths (*Annulithus arkelli*), indicating the *A. arkelli* Biozone of Hamilton (1982), were reported (Hamilton, 1982; but see Bown, 1987). Lord and Boomer (1990) reported ostracods, including taxa indicating the *Ogmoconchella aspinata* Biozone of Boomer (1991). The presence of foraminifers of the *Glomospira/Glomospirella* Assemblage was reported by Copestake (1989), and scolecodonts were recorded by Warrington (1981) and Warrington and Whittaker (1984).

The lithological sequences recorded in the Selworthy No. 2 Borehole [SS 9244 4618], together with that recorded by Richardson (1911) from Blue Anchor Point and the nearby foreshore section [ST 040 438], and that given in Whittaker and Green (1983) from a locality [ST 0765 4362] on the foreshore just east of Watchet Harbour, are illustrated in Figure 27.

The formation is about 13.9 m thick at Watchet, about 14.1 m at Blue Anchor, and 10.28 m (corrected) in Selworthy No.2 Borehole.

Mayall (1979) reported a significant change in clay mineralogy at the Blue Anchor Formation–Westbury Formation junction at St Audries's Bay; the proportion of illite–smectite is higher in the Westbury Formation, and vermiculite appears. These changes are probably related to derivation of detrital minerals from a surrounding landmass under increasingly humid conditions.

The diagenetic modification of primary sedimentary fabrics in the Westbury Formation at St Audrie's Bay was discussed by Macquaker (1984), and the organic geochemistry of black shales in the formation at Watchet has been described by Macquaker et al. (1986), Thomas et al. (1993), and Tuweni and Tyson (1994).

## Lilstock Formation

The formation is about 3 m thick at St Audries's Bay, where the lower (Cotham) and upper (Langport) members are approximately equal in thickness. A thickness of 2.3 m was recorded at Blue Anchor (Richardson, 1911). In Selworthy No. 2 Borehole the corrected thickness of the formation, penetrated between 24.77 and 27.02 m, is 2.18 m (Figure 27).

The **Cotham Member** consists of pale grey and greenish grey calcareous mudstones and thin pale green limestones and sandstones; it is broadly fining-upwards in character. A thickness of up to 1.4 m was recorded in Doniford Bay [ST 0827 43545], just east of Watchet (Whittaker and Green, 1983). Richardson (1911) recorded 5

**Figure 28**
Stratigraphical distribution and relative abundances of palynomorphs in the Blue Anchor Formation, Penarth Group and basal Lias Group of Selworthy No. 2 Borehole. Relative abundances expressed as percentages based upon counts of 200 specimens. After Warrington et al., 1995, fig. 3.

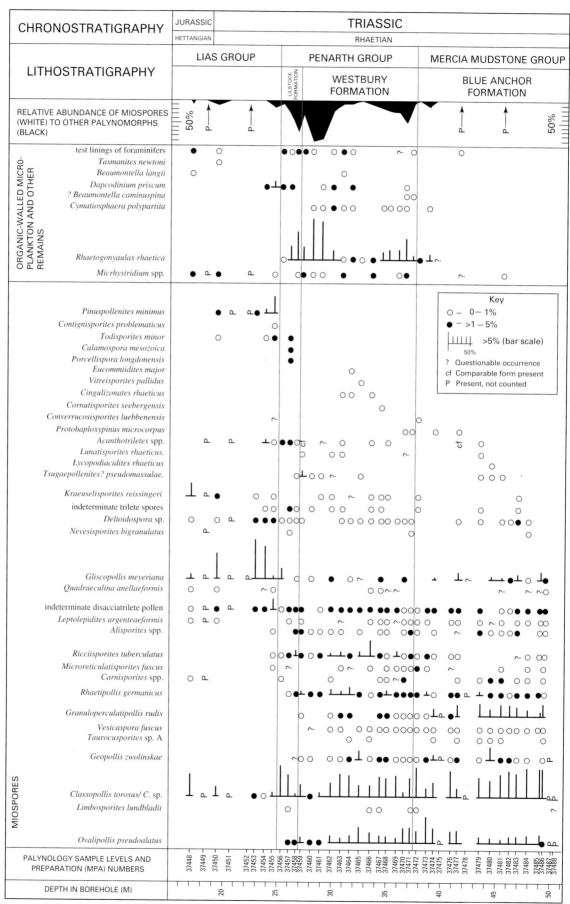

feet 4 inches (1.62 m) on the foreshore near Blue Anchor Cliff; in Selworthy No. 2 Borehole the member was about 1 m thick, between 26.03 and 27.02 m depth.

The junction of the member with the underlying West-bury Formation is locally sharp and may be an erosional non-sequence; however in Selworthy No. 2 Borehole, it appears to be transitional. In the coastal exposures the base of the member is commonly faulted (Whittaker, 1978b).

The lower part of the member contains a compara-tively thin unit (up to about 0.5 m) of greenish grey silt-stones and marly shales with contorted, deformed and possibly slumped calcareous siltstones. The top of the contorted beds is commonly dissected by a crudely poly-gonal network of deep shrinkage cracks caused by expo-sure and desiccation (compare with Whittaker and Green, 1983, pl.10). In section these cracks are seen as narrow vertical sheets filled with calcarenite which may also form a thin layer linking the tops of the fissures. In the coastal exposures, the thin layer of coarse-grained crack infill is succeeded by a further sequence of mainly greenish grey shales, locally faintly laminated, with silt-stone wisps and some harder calcareous lenses and nodules. Whittaker (1978b) established that the Cotham Marble is not represented in these sequences, the beds identified as such by Richardson (1911) being the basal bed of the Langport Member.

The basal bed of the Cotham Member in Selworthy No. 2 Borehole is a greenish grey, fine-grained sandstone (about 26.67 to 27.02 m depth); the junction with the underlying Westbury Formation is transitional. At 26.39 m depth, a pale grey limestone with calcite veins contains deep burrow-like fills with coarse sand-sized calcareous peloids comparable with the infill of the shrinkage-cracks horizon at outcrop (Bed 3 of the Cotham Member; Whit-taker and Green, 1983). In the borehole, the underlying beds are very pale grey to greenish grey mudstone over-lying a thin green-grey sandstone to 27.02 m depth.

No bivalves were observed in the member in the bore-hole. Palynomorph assemblages from the Cotham Member (Figure 28) are similar to those in the upper part of the Westbury Formation, and include a high pro-portion of dinoflagellate cysts (*Rhaetogonyaulax rhaetica*).

At outcrop in the eastern part of the district the Cot-ham Member yielded a sparse bivalve fauna from the base (Richardson, 1911), but it is generally nearly barren of macrofauna other than scattered fish remains. Micro-fossils recorded nearby include spores, pollen, acritarchs, dinoflagellate cysts and tasmanitid algae (Warrington, 1974, 1981, 1983); Hamilton (1982) reported the pres-ence of coccoliths, including *Schizosphaerella punctulata* indicating the *S. punculata* Biozone of Bown (1987). Ostracods (Mayall, 1983; Lord and Boomer, 1990) and sparse foraminifera of the *Glomospira/Glomospirella* Assem-blage (Copestake, 1989) also occur.

The **Langport Member** comprises three or four beds of limestone divided by interbeds of grey or blue-grey mud-stones, and is only up to about 1 m thick in the district. The lower limestones are commonly lenticular or nodular, porcellanous or laminated. The higher group of two or three limestone beds weathers to a distinctive creamy white colour and has an irregular base overlain locally by pebbles. The top is also irregular in places and is pene-trated locally by U-shaped burrows. The uppermost lime-stone is termed the 'Sun Bed'; above it are grey calcareous mudstones with limestone lenticles; this unit is about 0.7 m thick near Watchet. It is succeeded, without a significant break, by grey shaly calcareous mudstones and shales of the earliest beds of the Lias Group (Whittaker, 1978b).

In Selworthy No. 2 Borehole, the Langport Member consists of 1.26 m of limestone and calcareous mudstone; its base is placed at the base of a limestone at 26.03 m (Figure 27). The uppermost limestone (24.77 to 25.17 m depth) is equivalent to the 'Sun Bed'; it is micritic, con-tains vugs and calcite veins, and is extensively sheared. It overlies pale grey, calcareous, silty mudstones, with a thin limestone interbed and a thicker basal limestone; at 25.68 m depth, a bed of greenish grey mudstone is rich in ostracods, but no other fauna was seen. Palynomorph assemblages from the member are slightly less diverse than those recorded lower in the borehole; they are dominated by miospores, mainly *Classopollis torosus*, *Glisco-pollis meyeriana* and bisaccate pollen, but include acrit-archs and dinoflagellate cysts (Figure 28).

At outcrop immediately to the east of the district the member has yielded few macrofossils; corals and a few bivalves have been recorded at St Audrie's Bay (Warring-ton and Ivimey-Cook, 1995). The member has yielded spores, pollen, acritarchs and dinoflagellate cysts (War-rington, 1974, 1981, 1983). Hamilton (1982) recorded coccoliths indicating the *Schizosphaerella punctulata* Bio-zone; also reported are foraminifera of the *Eoguttulina liassica* Assemblage (Copestake, 1989) and ostracods (Lord and Boomer, 1990).

## Depositional environments

The depositional conditions of the Penarth Group have to be considered in the context of the change from the largely continental environment represented by the red Mercia Mudstone to the epicontinental marine environ-ment represented by the Lias Group. The onset of a marine transgression from the Tethyan province to the south is first reflected in the organic geochemistry and biota of the upper beds of the Rydon Member of the Blue Anchor Formation at Watchet (Thomas et al., 1993). The faunas and microfloras of the succeeding Penarth Group reflect the progress of the transgression and the colonisation of the new marine environments. After some fluctuations, principally during deposition of the Cotham Member, the marine environment and the associated biota stabilised.

The Westbury Formation may comprise at least three sedimentary cycles which represent deposition in trans-gressive, littoral, high-energy environments, with local emergence, and lower energy, stagnant or weakly oxy-genated water bodies (Whittaker and Green, 1983).

The origin of the contorted beds and the desiccation cracks in the Cotham Member was reviewed by Mayall (1983) who concluded that the deformation was prob-ably caused by earthquake activity, and that minor uplift produced conditions suitable for surface desiccation.

# SIX

# Jurassic

Jurassic rocks underlie most of the offshore area, where they occupy an east–west-trending, westward-plunging, faulted synclinal structure, forming part of the Bristol Channel Basin (Figure 29). Most of the eastern half of the submarine outcrop consists of Early Jurassic strata, while Mid and, possibly, Late Jurassic strata are preserved to the west in the core of the syncline. Early Jurassic strata (Lias Group) are also present onshore, between Blue Anchor [ST 035 434] and Watchet [ST 070 436], and in the Porlock Basin near Selworthy [SS 918 467] (Figure 29). The lowermost 6 m or so of the Lias Group (the 'Pre-planorbis Beds') are of Triassic age, but for the purpose of this account are described in this chapter. The stratigraphy of the onshore area (which, although

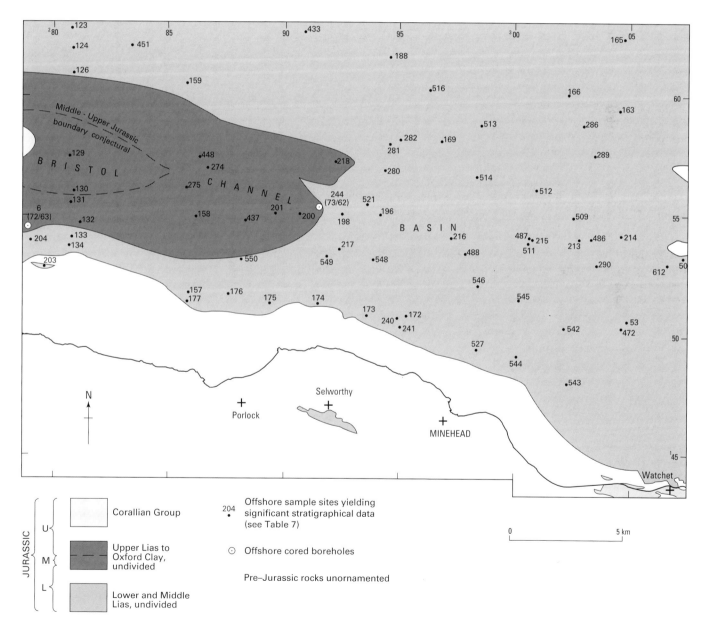

**Figure 29** Distribution of Jurassic rocks in the district, with locations of offshore samples and offshore cored boreholes. Faults and drift deposits omitted.

less complete, is known in greater detail) is discussed before that of the offshore area.

## ONSHORE AREA

### Lias Group

The most extensive outcrop is found in the cliffs and foreshore between Blue Anchor and Watchet where there is excellent exposure. However, the sequence is folded and much broken up by faults so that nowhere is there an uninterrupted section through the whole succession. About 10 km west of the main coastal outcrop, a small (about 2.5 km by up to 0.8 km) outlier, the most westerly known representative of Jurassic rocks in England, occurs near Selworthy (Figure 29; Whittaker, 1976; Warrington et al., 1995). The outcrop trends east-south-east, and dips are to the north at 16–20° (Figure 20). The northern margin is faulted against the Devonian Hangman Sandstone; on the south side, there is a downward passage into the Penarth Group. As on the coast, the outcrop is affected by numerous faults, only a small number of which are shown on the map.

The main lithologies are pale, medium and dark grey calcareous mudstones and shales, commonly fissile and locally bituminous but in places poorly laminated or blocky, alternating with pale grey, micritic, locally shelly, limestones. This 'limestone-shale' alternation is an over-simplification of a more complex cyclic pattern of basal shale (commonly with a sharp base)–mudstone–limestone–mudstone. During weathering, the 'paper-shale' appearance of some shales may be enhanced by the development of selenite on the bedding planes. The limestones, commonly with sharp bases, may be laminated, micritic, and locally nodular. Individual limestone beds, and combinations of beds, can be traced for considerable distances along the coast; however, the sharp contacts apparent between the limestones and the adjacent argillaceous rocks at outcrop are not seen where the effects of weathering are absent, as in borehole cores. Fossils are commonly concentrated about some of these junctions which may therefore represent more extended periods of slower deposition. Of the mudstones, the paler grey types are more commonly burrowed and show trace fossil mottling. These lithologies give rise to yellowish brown clayey soils, locally with scattered limestones.

The origin of the limestone/shale rhythms has been the source of much discussion (e.g. Hallam, 1964). Whittaker and Green (1983, p.60) noted that the wide extent of individual limestones (but with great variation in the thicknesses of the intervening mudstones), together with the presence of trace fossil mottling and the regular pattern of the lithological cycle, favoured a primary or very early diagenetic origin for most of the limestone beds (although some of the nodular limestones might be secondary). Weedon (1986) concluded that the carbonate mud was supplied by coccoliths in zooplankton faecal pellets (now largely neomorphosed to microspar), resulting in hemipelagic sedimentation with the clay fraction being supplied from land sources. He related the cyclicity

apparent in the sediments to climatic changes resulting from orbital changes in insolation, but also recognised the effects of severe storm events. Carbonaceous laminae in the fissile bituminous mudstones may have resulted from the repeated accumulation of organic matter at or closely below the sediment/water interface in anaerobic conditions.

The thicknesses and lithologies of the beds exposed in the foreshore are closely comparable with those described by Whittaker and Green (1983) from the adjacent Weston-super-Mare district (Sheet 279) where they divided the Lias Group into 257 numbered beds, grouped together into five mappable lithological divisions, summarised here in Figure 30. Beds at least as high as 210 are exposed in the district and, following the terminology used in the Weston-super-Mare district, are all assigned to the Blue Lias Formation, with an estimated total thickness of about 150 m. In that district, the top of the Blue Lias was taken at the top of Bed 250, that is at the top of the highest recorded limestone, but it will almost certainly be lowered when a lithostratigraphical definition, applicable in all areas, is agreed.

Vertical 1:5000 scale aerial photographs of the foreshore area between Blue Anchor and Watchet, flown in 1967, enabled detailed mapping of the structure and, in conjunction with examination of the exposures, allowed the five lithological divisions to be mapped out (Figure 31; Whittaker, 1972).

North of Blue Anchor Point, the Blue Lias outcrop [ST 0382 4394 to 0441 4378] occupies a syncline faulted along the northern side. Most of the western part is Division 1 but eastwards [east of ST 0410 4384], divisions 2 and 3 are represented in faulted outcrops. Sections [ST 0408 4378] measured by Dr A Whittaker in 1970 showed the lowest part of Division 1 separated by a gap from a sequence in the upper part of Division 1 and the basal part of Division 2. A fault-bounded outcrop [ST 0444 4393], 550 m east-north-east of Blue Anchor Point (Figure 31) and surrounded by Mercia Mudstone, is mainly Division 3 with some Division 4 and Division 1 on the south-west side.

East of Blue Anchor, just south of a headland [ST 0479 4366] of red Mercia Mudstone, and probably separated from it by an east-west fault, slipped masses of grey to dark grey shale with large yellow-brown-weathering limestone blocks form low cliffs. On the foreshore, these Division 5 shales are faulted along their northern margin against Division 4, and then extend for almost a kilometre eastwards, in a complex faulted outcrop, to the Watchet Fault (Figure 31). At the base of the cliff, and separated from it by a fault, they form reefs dipping south at 60°. In the cliff [ST 0514 4348], extensive bedding surfaces dip north

**Figure 30** Divisions in the Penarth Group and Lias Group of the district, showing lithologies, ammonite zonation, and typical fossils; bed numbers and divisions of the Lias Group after Whittaker and Green (1983); illustrations based on publications of the Natural History Museum, London.

| SYSTEM | LITHO-STRATI-GRAPHY | | LITHOLOGY | BED NUMBER | STAGE | AMMONITE BIOZONATION: ZONE | AMMONITE BIOZONATION: SUBZONE |
|---|---|---|---|---|---|---|---|
| JURASSIC | LIAS GROUP | 'DIVISION 5' | Mudstones and shales with a few argillaceous limestones | 204 - 257 | SINEMURIAN | *Arnioceras semicostatum* (BED 209) | *Euagassiceras resupinatum*; *Agassiceras scipionianum*; *Coroniceras lyra* |
| JURASSIC | LIAS GROUP | 'DIVISION 4' | Fissile shales and mudstones with a few limestones in middle and upper parts; alternating shales and massive limestones in lower part | 147 - 203 | SINEMURIAN | *Arietites bucklandi* (?BED 208 – BED 146) | *Coroniceras rotiforme*; *Coroniceras conybeari* |
| JURASSIC | LIAS GROUP | 'DIVISION 3' | Alternating, commonly nodular, limestones and shales; individual limestones become thicker and more massive towards top | 69-146 | HETTANGIAN | *Schlotheimia angulata* (BED 145 – BED 80) | *Schlotheimia complanata*; *Schlotheimia extranodosa* |
| JURASSIC | LIAS GROUP | 'DIVISION 2' | Dark shales and mudstones with a few nodular limestone horizons | 40 - 68 | HETTANGIAN | *Alsatites liasicus* (BED79 – BED 43) | *Laqueoceras laqueus*; *Waehneroceras portlocki* |
| JURASSIC | LIAS GROUP | 'DIVISION 1' | Alternating thin limestones and calcareous shales passing down into calcareous shales | 1 - 39 | HETTANGIAN | *Psiloceras planorbis* (BED 42 – BED 8) | *Caloceras johnstoni*; *Psiloceras planorbis* |
| TRIASSIC | LIAS GROUP | | 'Pre-planorbis Beds' | | | BED 7 | |
| TRIASSIC | PENARTH GROUP: LILSTOCK FORMATION | LANGPORT MEMBER | 'SUN BED' Thin, pale grey limestones and grey mudstones | | | | |
| TRIASSIC | PENARTH GROUP: LILSTOCK FORMATION | COTHAM MEMBER | Limestones and silty calcareous mudstones | | | | |
| TRIASSIC | PENARTH GROUP: WESTBURY FORMATION | | Mudstones, silty mudstones, thin limestones and thin sandstones | | | | |

**SOME CHARACTERISTIC FOSSILS**

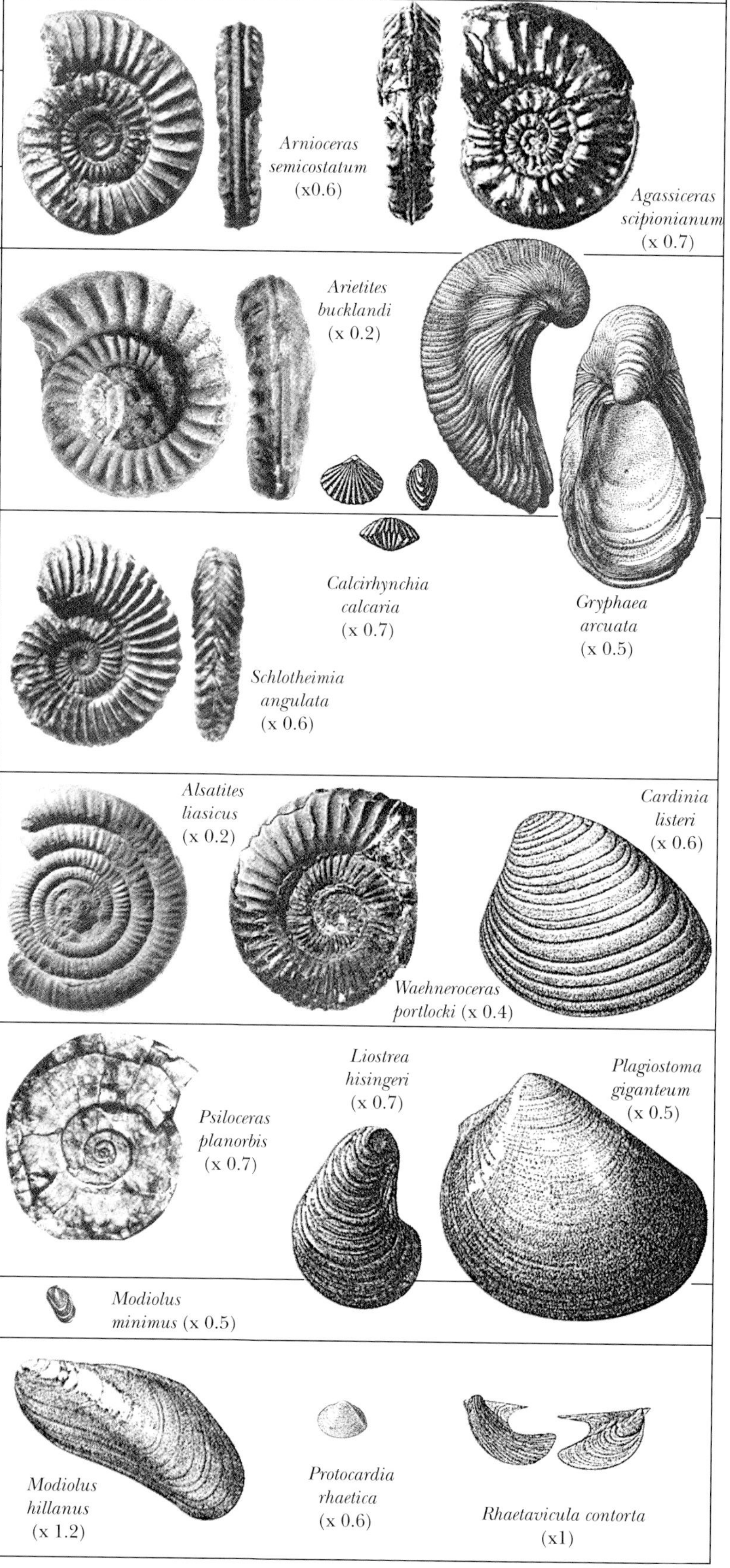

Arnioceras semicostatum (x0.6)
Agassiceras scipionianum (x 0.7)
Arietites bucklandi (x 0.2)
Calcirhynchia calcaria (x 0.7)
Gryphaea arcuata (x 0.5)
Schlotheimia angulata (x 0.6)
Alsatites liasicus (x 0.2)
Cardinia listeri (x 0.6)
Waehneroceras portlocki (x 0.4)
Liostrea hisingeri (x 0.7)
Plagiostoma giganteum (x 0.5)
Psiloceras planorbis (x 0.7)
Modiolus minimus (x 0.5)
Modiolus hillanus (x 1.2)
Protocardia rhaetica (x 0.6)
Rhaetavicula contorta (x1)

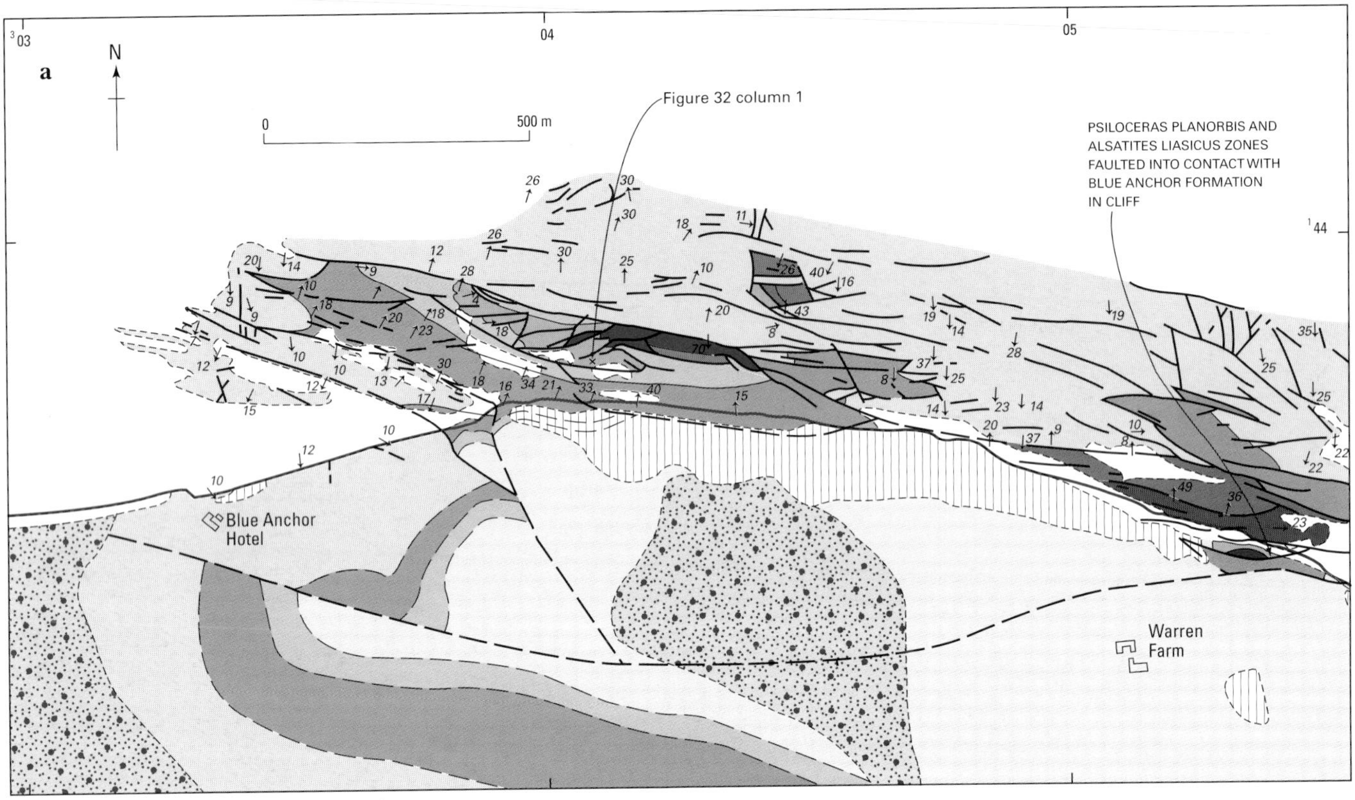

03  N  04  05

a

0      500 m

Figure 32 column 1

PSILOCERAS PLANORBIS AND
ALSATITES LIASICUS ZONES
FAULTED INTO CONTACT WITH
BLUE ANCHOR FORMATION
IN CLIFF

44

Blue Anchor
Hotel

Warren
Farm

**Figure 31a and b**   Geological map of foreshore and cliff sections in Late Triassic and Early Jurassic rocks between Blue Anchor and Watchet.

at about 50°. Slipping has occurred on the seaward-dipping bedding surfaces.

At a small headland [ST 0536 4343], Blue Anchor Formation is faulted against Blue Lias (*planorbis* and *liasicus* zones) (Plate 23). Eastwards, the Blue Anchor Formation forms the lower part of the cliff [at and west of ST 0548 4338] and is again faulted, in the upper part, against Blue Lias. Farther east, Mercia Mudstone ('Variegated Marls') is faulted against a similar level of the Blue Lias at a headland [ST 0552 4338], south of which the latter formation forms the cliff. The beds dip at 10° to 050°, and consist of about 6 m of dark grey mudstone and shale with regular 0.1 to 0.2 m thick interbeds of grey limestone.

Low cliffs [between ST 0553 4335 and ST 0572 4332] are mainly slipped Division 2. Grey to dark grey shales with few limestones are seen at beach level at the toe of slipped areas. Eastwards [ST 0577 4333], the Blue Lias dips at 22° to 267°, and consists predominantly of grey shales and mudstones with few limestones (lower part of

**Plate 23**   Small headland [ST 0536 4343] about 300 m north-east of Warren Farm showing Lias Group (*planorbis* and *liasicus* zones) on right faulted against Blue Anchor Formation (left-hand side) (1972) (A 11712).

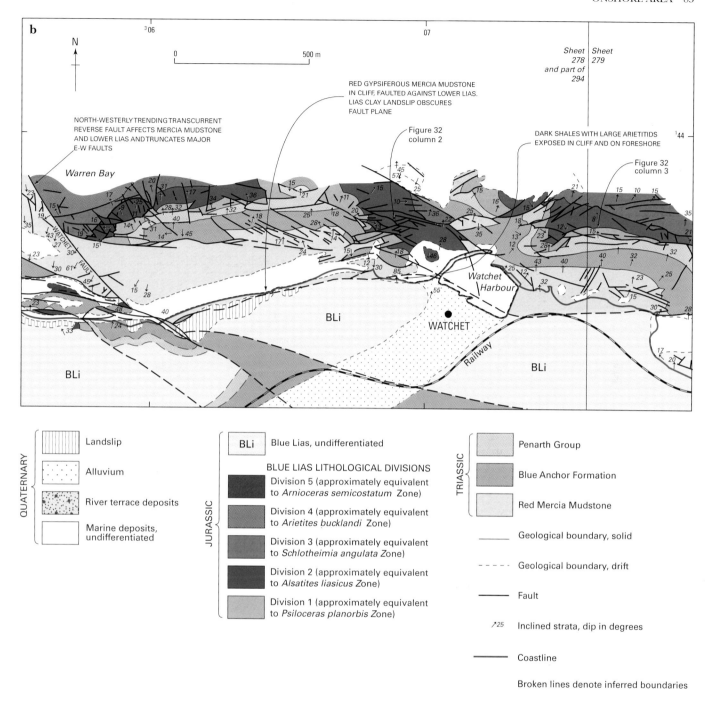

Division 1), of which about 12 m are exposed. This sequence is overlain by shales with prominent yellow-brown-weathering limestones up to 0.3 m thick. At beach level, just west of the Watchet Fault, a fault-bounded outcrop of shales with some limestones shows dips of 70° to 175°.

In Warren Bay, about 400 m north of the coastline, and east of the Watchet Fault, reefs show Blue Lias on the north side of a faulted anticline (Figure 31). Lithological divisions 1 to 4 are present in the faulted outcrops. Between the Watchet Fault [ST 0607 4332] and West Street Beach [ST 0683 4349], a major fault trending between east and east-north-east brings red Mercia Mud-

stone in the lower cliff against Blue Lias in the upper cliff (Figure 31); slips are present along much of the cliff. The fault plane [ST 0616 4338] is estimated to dip at about 60° south.

Farther east [ST 0635 4342], partly slipped grey to dark grey shales with limestones, estimated to dip north at about 25°, are present in the upper cliff; the fault plane between these Blue Lias beds and the gypsiferous Mercia Mudstone in the lower cliff is estimated to dip south at about 20°. In an area of recent slips [ST 0650 4344], the Blue Lias is estimated to dip north at 50–60° (locally steeper), and the fault plane between the Blue Lias and the Mercia Mudstone is estimated to dip south at about

40°. Inaccessible Blue Lias seen in the upper cliff at about [ST 0660 4348], apparently consists of up to 20 m of grey shale with six limestone beds 0.1 to 0.3 m thick. Whittaker (MS) recorded an 85° dip to the north [at ST 0689 4350]; somewhat shallower (50–60°) northerly dips were recorded near the western end of the slipway [ST 0686 4349].

Between about 100 and 400 m north of West Street Beach, the foreshore shows much-faulted outcrops of divisions 1 and 2 which extend into the harbour area (Figure 31). A section [around ST 069 437] west of the harbour is shown graphically in Figure 32. From a site [ST 0739 4362] 150 m north-east of the harbour entrance, exposures of divisions 1 to 3 form part of a larger outcrop that extends eastwards out of the district into Doniford Bay. A measured section [ST 0765 4367] in Division 1, recorded on the foreshore about 500 m east of the harbour, is shown graphically in Figure 32.

Divisions 1 and 2 were recovered in Selworthy No. 2 Borehole [SS 9244 4618], and badly fractured cores in alternating limestones and mudstones of Division 3 were recovered in Selworthy No. 1 Borehole [SS 9244 4630] (Figure 33; Warrington et al., 1995).

The Blue Lias is in conformable contact with the underlying Penarth Group (Langport Member of the Lilstock Formation); there is a quite rapid transition from the micritic limestones of the 'Sun Bed', through a small thickness of marly beds, into a few metres of calcareous mudstones and shales with very few limestones. The last-named unit was formerly assigned to the 'Watchet Beds' of the 'Upper Rhaetic' by Richardson (1911), who equated it with a similar thickness of grey, somewhat fissile, silty, calcareous shales on the Glamorgan coast. However, Whittaker (1978 b) concluded that, in Somerset, these beds were generally better assigned to the Lias Group, and drew the base of that group (and the Blue Lias Formation) at the top of the marly beds just above the 'Sun Bed'.

The faunas of the different lithologies vary considerably. Ammonites are sporadically abundant in the shales, where they are generally crushed flat, whilst in the limestones they appear to be scarcer but may be partly infilled and preserved uncrushed. In the lower beds, they are commonly associated with low-diversity bivalve faunas and fish remains. The more bituminous shales contain virtually no benthic macrofauna and lack signs of burrowing activity, but some horizons contain crushed ammonites. The paler grey mudstones have only a limited benthic fauna with some shallow-burrowers, but nektonic faunas comprising ammonites and some of the more motile bivalves are more common. In the limestones, a low-diversity macrofauna may be present, commonly concentrated on the upper surfaces; this may include the bivalves *Gryphaea*, *Liostrea* and *Plagiostoma*, and a limited number of nektonic taxa including pectinid bivalves and ammonites. The small brachiopod *Calcirhynchia* is found scattered, or in small clusters, within some micritic limestones.

The ammonites provide the primary biostratigraphical control for these beds; the standard ammonite biozonation, which can be used as the basis of a chronozonation, is shown in Figure 30. Following Whittaker (1972), the five mappable lithological divisions mentioned above approximate to the three Hettangian and two oldest Sinemurian ammonite zones.

The strata between the top of the Penarth Group and the first appearance of the ammonite *Psiloceras* are, by definition, Triassic in age (Cope et al., 1980; Warrington et al., 1980), and are commonly referred to as the 'Pre-planorbis Beds' (= lower part of Division 1); Selworthy No. 2 Borehole proved about 6.3 m (Figure 33; Warrington et al., 1995). In 1970, Whittaker (MS) recorded thicknesses of at least 6 m in a section [ST 0408 4378] east of Blue Anchor Cliff (Figure 32) and 8.3 m in a section [ST 0765 4367] on the foreshore east of Watchet Harbour, about 50 m east of the district boundary (Figure 32). At St Audrie's Bay in the adjoining Weston-super-Mare district, the 'Pre-planorbis Beds' now equate with beds 1–7 of Whittaker and Green (1983), following the discovery of *Psiloceras* in Bed 8 (Hodges, 1994); this section is of particular importance because it is proposed as a possible Global Stratotype Section and Point for the base of the Jurassic System (Warrington et al., 1994).

The fauna of the 'Pre-planorbis Beds' is dominated by the oyster *Liostrea hisingeri* and small species of the mussel *Modiolus*, (e.g. *M. minimus*). The presence of *L. hisingeri* covering the upper surfaces of some limestone beds has led some authors (e.g. Bristow and Etheridge, 1873) to use the name 'Ostrea Beds' for this unit. Other bivalve genera recorded in the generally sparse macrofauna include *Chlamys*, *Palaeoneilo*, *Plagiostoma*, *Protocardia* and *Pteromya*. A marker bed with common *Plagiostoma giganteum* (= Bed 7f at St Audrie's Bay) is traceable throughout Somerset, Dorset and South Wales (Hodges, 1994). A section [SS 9179 4622] in Great Wood, Selworthy (Figure 20), showed 1.74 m of 'Pre-planorbis Beds' with bivalves including *Liostrea hisingeri*, *Modiolus minimus* and *Pteromya tatei*, as well as diademopsid echinoid spines. Palynomorphs from these beds in Selworthy No. 2 Borehole are dominated by circumpolles (e.g. *Gliscopollis meyeriana*) and are much less diverse than those in the underlying Penarth Group (Figure 28).

The base of the *Psiloceras planorbis* Zone (and the Jurassic System) is placed at the base of Bed 8 at St Audrie's Bay (Hodges, 1994) where the zonal thickness is 8.57 m; in Selworthy No. 2 Borehole it is about 10 m thick (Figure 33; Warrington et al., 1995). The ammonite *Psiloceras planorbis* is locally abundant as compressed and nacreous specimens (e.g. in overhanging bedding surfaces at the small headland [ST 0536 4343] and in large fallen blocks on the beach [ST 0616 4338]), and *Caloceras johnstoni* —the index species for the younger subzone of the *planorbis* Zone — is recorded by Whittaker (MS) in foreshore sections [ST 0408 4378] about 200 m east of Blue Anchor Cliff. Other elements of the sparse macrofauna recorded hereabouts include the bivalves *Anningella*, *Camptonectes*, *Liostrea*, *Protocardia* and *Pteromya*, as well as echinoid remains and fish fragments. Both subzones were proved in Selworthy No. 2 Borehole (Figure 33), and specimens of *Psiloceras planorbis* were recorded in an old pit [SS 9246 4612] at Pixie, and in old workings [SS 9256 4613] along strike from there (Figure 20). A palynomorph assemblage from the basal part of the Zone

**Figure 32** Measured sections of the Lias Group between Blue Anchor and Watchet; locations shown on Figure 31.

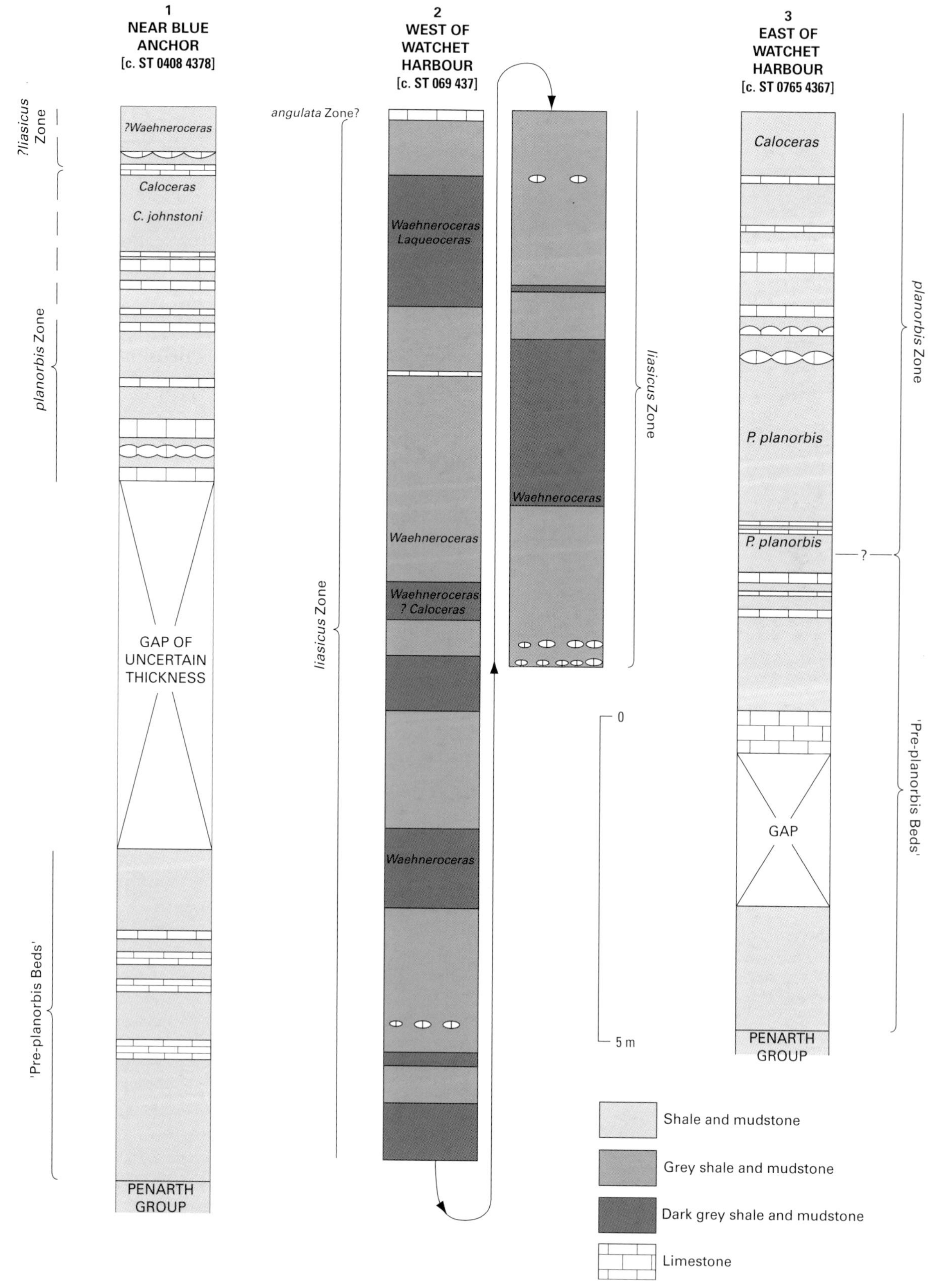

Column 1: Foreshore exposures in 'Pre-planorbis Beds' and *planorbis* and ?lowest *liasicus* zones about 260 m ENE of Blue Anchor Cliff.

Column 2: Foreshore exposures in *liasicus* Zone strata just west of Watchet Harbour.

Column 3: Foreshore exposures in 'Pre-planorbis Beds' and *planorbis* Zone strata about 500 m east of Watchet Harbour (50 m east of the district boundary).

All sections measured by Dr A Whittaker in 1970.

**Figure 33**
Graphic logs of
the Selworthy
boreholes, with
distribution of
ammonites
recorded from
No. 2 Borehole
(after Warrington
et al., 1995).

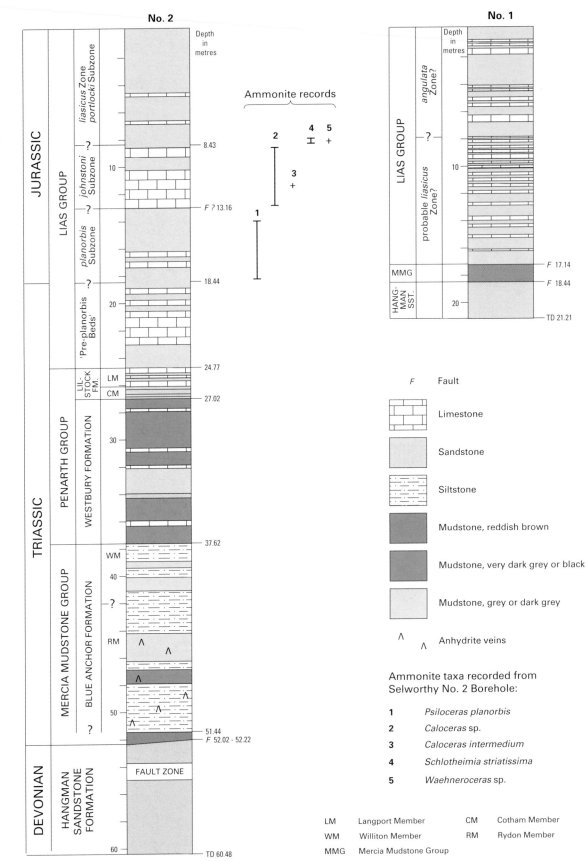

Ammonite records

*F*    Fault

Limestone

Sandstone

Siltstone

Mudstone, reddish brown

Mudstone, very dark grey or black

Mudstone, grey or dark grey

∧   ∧   Anhydrite veins

Ammonite taxa recorded from
Selworthy No. 2 Borehole:

1    *Psiloceras planorbis*

2    *Caloceras* sp.

3    *Caloceras intermedium*

4    *Schlotheimia striatissima*

5    *Waehneroceras* sp.

| | | | |
|---|---|---|---|
| LM | Langport Member | CM | Cotham Member |
| WM | Williton Member | RM | Rydon Member |
| MMG | Mercia Mudstone Group | | |

in Selworthy No. 2 Borehole (Figure 28) is comparable with that from the 'Pre-planorbis Beds'.

In west and central Somerset, the overlying *Alsatites liasicus* Zone is about 27 m thick (Ivimey-Cook and Donovan, 1983). A measured section [ST 069 437] just west of Watchet Harbour showed the full thickness of the zone (Figure 32). It is characterised by species of the ammonite *Waehneroceras*. The zonal index genus *Alsatites* is very scarce (Ivimey-Cook and Donovan, 1983). A number of finely ribbed juvenile *Schlotheimia striatissima* were recovered from the lowest beds in Selworthy No. 2 Borehole (Warrington et al. 1995). *Waehneroceras* and *Laqueoceras* (indicating the upper part of the *liasicus* Zone) were noted by Whittaker (MS) on the foreshore [ST 0670 4365]. The non-ammonite faunas are more diverse than in the underlying and overlying zones. The coral *Heterastraea* has been recorded hereabouts, and echinoid and fish fragments occur with the bivalves *Camptonectes*, *Gervillella*, *Gryphaea*, *Lucina*, *Liostrea*, *Modiolus*, *Plagiostoma* and *Pseudolimea*.

The base of the *Schlotheimia angulata* Zone is recognised by the appearance of the nominal taxon (Figure 30). Just east of the district, in Doniford Bay, the zone is up to 41.7 m thick (Whittaker and Green, 1983, pp.63–64). Large fallen blocks of shale with abundant *Schlotheimia* recorded on the beach [ST 0643 4346] by Whittaker (MS) indicate that the zone is present in the upper cliff. Near Selworthy (Figure 20), a few loose blocks of dark bluish grey, argillaceous limestone in an overgrown pit [SS 9216 4632] (= 'Middle Pit') included casts of a strongly ribbed ammonite now identified as *Schlotheimia* cf. *extranodosa* from the lower part of the *angulata* Zone (contrary to Whittaker's (1976) comment that the specimen indicated the higher part of that Zone). The non-ammonite fauna is rather sparse and not well preserved. *Gryphaea arcuata* and the brachiopod *Calcirhynchia calcaria* both become increasingly abundant towards the top. The bivalves *Anningella*, *Camptonectes*, *Cardinia* and *Plagiostoma*, and the gastropod *Amberleya*, together with fish scales and fragments of fish and reptile bone, have also been recorded hereabouts.

The appearance of arietitid ammonites is taken to indicate both the base of the *Arietites bucklandi* Zone and the base of the Sinemurian Stage. Large arietitids were recorded by Whittaker (MS) [ST 0693 4349], just north of the slipway at West Street Beach, close to the east–west fault that brings the Blue Lias against red Mercia Mudstone. The *bucklandi* Zone is about 50 m thick hereabouts, and in its lower part (*Coroniceras conybeari* Subzone) contains numerous specimens of *Coroniceras*. The overlying *Coroniceras rotiforme* Subzone is characterised by *Coroniceras rotiforme* and *C. hyatti*. The non-ammonite fauna is relatively sparse; the brachiopod *Calcirhynchia calcaria* is locally abundant, and bivalves including *Camptonectes*, *Meleagrinella*, *Oxytoma* and *Semuridia*, with some crinoid columnals, fish scales, bone fragments and wood, have been recorded nearby. *Calcirhynchia calcaria* was recorded from loose blocks of typical Blue Lias limestone in an old pit [SS 9243 4633] (= 'Top Pit') near Selworthy (Figure 20).

The *Arnioceras semicostatum* Zone is the youngest unit present in the coastal sections of the district; the base is marked by the appearance of the ammonite *Arnioceras* which is relatively common in the fissile mudstones; coroniceratids dominate in the limestones. Whittaker (MS) recorded *Arnioceras* in low cliffs just south of the headland [ST 0479 4366]. The non-ammonite macrofauna of the zone is varied and locally abundant hereabouts; it includes serpulids, terebratuloid brachiopods, bivalves (*Anningella*, *Camptonectes*, *Chlamys*, *Gryphaea*, *Isognomon*, *Lucina*, *Meleagrinella*, *Modiolus*, *Myoconcha*, *Oxytoma*, *Pinna* and *Plagiostoma*), gastropods (including *Pleurotomaria*), fish, and reptile remains.

The faunas provide some indication of the changing depositional conditions during these latest Triassic and early Jurassic times. The presence of abundant *Liostrea*, *Modiolus* and echinoid spines in the 'Pre-planorbis Beds' suggests the existence of shallow marine water. The succeeding limestones and shales contain a deeper-water bivalve fauna, including *Plagiostoma giganteum*. This change indicates an expanding marine transgression which resulted in a shallow continental (epeiric) sea over most of southern Britain, prior to the first appearance of the ammonite *Psiloceras*. Marginal facies of conglomerates and calcarenites are present in Glamorgan, indicating the proximity of a shoreline.

## OFFSHORE AREA

The Jurassic stratigraphy of the central Bristol Channel was described by Evans and Thompson (1979), based on the results of shallow seismic reflection, sonar and coring surveys. They divided the succession into 'Lower & ?Middle Lias', of Hettangian to Pliensbachian age; 'Upper Lias', of Toarcian age; 'Fuller's Earth Clay', of possible Bajocian and Bathonian age; 'Oxford Clay', of Callovian to early Oxfordian age; and 'Corallian Sands', of Oxfordian age. Although up to about 1500 m of this sequence are now estimated to occur within the district, it has not proved possible to recognise all the subdivisions because much of the succession is made up of lithologically similar argillaceous strata, and there is only a sparse coverage of sample points. Furthermore, the offshore area (Tappin et al., 1994) is structurally complex, with strata folded into a series of open and locally tight folds, and traversed by numerous faults of uncertain throw and orientation.

Details of microfossil taxa recovered from the sediment samples, which provide key biostratigraphical control, were given by Evans (unpublished PhD thesis, 1973), but the original samples and residues do not appear to have been retained. It has not therefore been possible to redetermine the faunas and reassess their stratigraphical significance in the light of more recent knowledge (see also Warrington and Owens, 1977; Wilkinson and Halliwell, 1980). The location of sample sites is shown in Figure 29, and those for which there is some stratigraphy are listed in Table 7. Only two cored boreholes have been drilled in the offshore area of the district (Figure 29); Borehole 73/62 [about SS 9173 5513] is located centrally and Borehole 72/63 [about SS 7880 5456] lies at the extreme western edge of the sheet area.

**Table 7**
Offshore sites on the Jurassic outcrop of the district for which there is usable stratigraphical information. They are ordered approximately east to west, and stratigraphically. The sites are numbered according to the BGS Coastal Geology database in which each number is preceded by 51/04. Site locations shown on Figure 29.

| Site | Stratigraphy | Lithology |
|---|---|---|
| 050 | Upper Sinemurian–Lower Pliensbachian | mudstone, grey |
| 612 | Hettangian–Sinemurian | mudstone, grey |
| 165 | Hettangian or Lower Sinemurian | limestone, grey, silty |
| 214 | *semicostatum* Zone (Lower Sinemurian) | clay, grey, shaly |
| 053 | ?Upper Pliensbachian | mudstone, grey, mottled khaki |
| 472 | Hettangian–Early Sinemurian | mudstone, grey |
| 486 | Hettangian–Sinemurian | siltstone, hard, grey |
| 286 | Late 'Trias'–Early 'Liassic' | limestone, grey |
| 166 | *turneri* Zone? (Lower Sinemurian) | marl, grey, micaceous |
| 509 | Sinemurian | mudstone, grey-green |
| 213 | ?*bucklandi* Zone (Lower Sinemurian) | clay, grey, shaly |
| 542 | Late Sinemurian–Pliensbachian | mudstone, grey to olive-green |
| 512 | Hettangian–Sinemurian | mudstone, green-grey |
| 215 | *angulata* Zone (Hettangian) | clay, grey |
| 511 | Early Sinemurian | siltstone, pale grey-green |
| 545 | Pliensbachian | mudstone, grey |
| 544 | Early Sinemurian | claystone, olive-grey |
| 513 | Hettangian–Sinemurian | clay, green-grey |
| 514 | Hettangian–Sinemurian | mudstone, grey |
| 546 | Hettangian–Sinemurian | limestone, grey |
| 527 | Early Sinemurian | mudstone, grey |
| 488 | Early Sinemurian | mudstone, dark grey |
| 216 | *angulata* Zone (Hettangian) | clay, grey, shaly |
| 169 | *raricostatum* Zone (Upper Sinemurian) | paper-shale, grey, micaceous |
| 516 | Rhaetian or Early Jurassic | mudstone, grey and red |
| 241 | 'Blue Lias' | limestone, hard, fine |
| 172 | Lower Sinemurian | clay, pale grey, micaceous, calcareous |
| 282 | Early Hettangian | mudstone, grey, shaly |
| 188 | Jurassic, '?Lower Lias' | limestone, grey, micaceous |
| 280 | Hettangian–Sinemurian? | mudstone, grey, shaly |
| 196 | *semicostatum* Zone (upper part) (Lower Sinemurian) | clay, grey, shaly |
| 240 | 'Blue Lias' | limestone, hard, fine |
| 173 | *angulata* or lower *bucklandi* Zone (upper Hettangian or Lower Sinemurian) | limestone, dark grey |
| 548 | Late Sinemurian | mudstone, grey |
| 521 | Late Sinemurian | mudstone, grey, shaly |
| 198 | 'Lower Lias?', *jamesoni* Zone (Lower Pliensbachian) | shale, grey, micaceous |
| 217 | *turneri* Zone (Lower Sinemurian) | clay, grey |
| 174 | near base *semicostatum* Zone (Lower Sinemurian) | limestone, grey, banded, fine |
| 549 | Late Sinemurian | mudstone, grey and pale olive-grey, mottled |
| 433 | Early Sinemurian | mudstone, grey and olive-green |
| 175 | *bucklandi* Zone (Lower Sinemurian) | clay, grey, calcareous |
| 550 | Pliensbachian | clay, stiff, grey, mottled |
| 176 | *semicostatum* Zone (Lower Sinemurian) | clay, grey, calcareous |
| 157 | *angulata* Zone (Hettangian) | mudstone, grey, shaly |
| 177 | Jurassic | clay, dark grey |
| 123 | 'Middle–Upper' Sinemurian | shale, grey |
| 134 | ?Hettangian | mudstone, dark grey |
| 133 | 'Lower Lias?' | mudstone, dark grey, shaly |
| 204 | *obtusum* Zone (Upper Sinemurian) | clay, yellow-grey |
| 203 | Lower Sinemurian | cementstone, grey, clayey |
| 451 | Lower Jurassic | mudstone, grey, silty |
| 281 | Lower Jurassic | mudstone, grey, shaly |
| 290 | Early Jurassic | mudstone, grey |
| 289 | Early Jurassic | mudstone, grey, shaly |
| 163 | 'Lias?' | shale, dark grey |
| 543 | Lias | mudstone, grey |
| 487 | 'Liassic' | siltstone, grey |
| 200 | 'Upper Lias', *falciferum* Zone? (Lower Toarcian) | shale, grey |
| 244* | Lower Toarcian (*falciferum–bifrons* zones) | mudstone, siltstone, very thin limestone and calcareous nodules (diagenetic compression of mudstones and drilling pressures have produced the appearance of clasts) |

**Table 7**
*continued.*

| Site | Stratigraphy | Lithology |
|------|-------------|-----------|
| 218 | Toarcian | shale, grey, micaceous |
| 201 | 'Upper Lias', *falciferum* Zone (Lower Toarcian) | clay, grey |
| 158 | Toarcian–Bathonian, '?Upper Lias' | mudstone, grey |
| 159 | Toarcian–Oxfordian | shale, brownish grey, with uncommon thin shell fragments |
| 437 | Earliest Aalenian | mudstone, grey |
| 448 | Aalenian–earliest Bajocian | mudstone, grey |
| 274 | Latest Bajocian | mudstone, pale grey |
| 275 | Latest Bajocian | mudstone, grey, shaly |
| 124 | 'Upper Lias'–'Oxford Clay' | shale, medium grey |
| 126 | 'Upper Lias'–'Oxford Clay' | shale, medium grey, with uncommon shell fragments |
| 132 | Toarcian–Bathonian | mudstone, grey, silty |
| 131 | Bathonian–Oxfordian | clay, pale grey and khaki-brown, mottled |
| 130 | Bathonian–Oxfordian | clay, soft, pale grey and khaki-brown, mottled, with common thin shell fragments, crinoid ossicles and belemnites |
| 129 | 'Upper Lias'–'Oxford Clay' | shale, hard, medium grey, with thin, hard, lenticular, sandy limestones with undulatory surfaces |
| 006† | Early Toarcian–Aalenian ('Upper Lias'–'basal Inferior Oolite') | mudstone, pale grey to greenish grey, with occasional shell fragments |

\* = Borehole 73/62
† = Borehole 72/63

## Lower and Middle Lias, undivided

The 'Lower & ?Middle Lias' subdivision of Evans and Thompson (1979) crops out on the sea bed over most of the eastern half of the map area (Figure 29). The top of this unit has been picked on deep seismic reflection profiles (see below). Where complete, in the westernmost part of the district, it is now considered to be up to 800 m thick. The lowest beds, of Hettangian and early Sinemurian age, consist of grey mudstones and shales with regular interbeds of thin, pale grey, micritic limestones, constituting the Blue Lias Formation as seen in the coastal exposures on both sides of the Bristol Channel. However, there are insufficient sample data to allow this formation to be shown separately on the map. Strata above the Blue Lias are grey mudstones and shales of late Sinemurian and Pliensbachian age.

## Upper Lias–Oxford Clay, undivided

The 'Upper Lias', 'Fuller's Earth Clay' and 'Oxford Clay' of Evans and Thompson (1979) are also not differentiated on the map which shows them as a single unit estimated to be up to about 700 m thick. This unit is predominantly argillaceous, with no indication of significant limestone developments corresponding to the Inferior and Great Oolite groups elsewhere. Its base is taken at an upward lithological change from grey mudstones and shales, to silty shales and mudstones. The lower boundary is evident on seismic profiles, suggesting the presence of a high-velocity interface, which possibly corresponds to the Marlstone Rock Formation elsewhere. The dating evidence demonstrates that the unit includes strata of Toarcian, Aalenian, Bajocian, Bathonian and Callovian age. It is equivalent to the Upper Lias, Inferior and Great Oolite groups, and Kellaways and Oxford Clay forma-

tions of the onshore Wessex Basin and elsewhere. In order to provide some stratal differentiation, a conjectural boundary between the Middle and Upper Jurassic (Callovian–Oxfordian stage boundary) is indicated (Figure 29). There is no unequivocal younger Jurassic sample known from the unit, but the presence of Late Jurassic (Oxfordian) mudstones is inferred from dated samples immediately west of the district boundary.

Within the unit, Evans and Thompson (1979) gave the thickness of 'Upper Lias', consisting of brown-grey mudstones and pale grey silty shales, as about 90 m in the central Bristol Channel. Thick sandstones typical of the Upper Lias onshore in Gloucestershire, Somerset and Dorset are lacking; the correlative beds show only silty horizons within shales. Evans and Thompson (1979) listed abundant foraminifera and ostracods, indicating a Toarcian age, together with bivalve fragments and belemnites. About 8 m of this interval were recovered in Borehole 73/62 which yielded a macrofossil assemblage including the ammonites *Dactylioceras*, *Hildaites* and *Hildoceras*, indicating the Lower Toarcian *Harpoceras falciferum* and *Hildoceras bifrons* zones (Ivimey-Cook, 1993). Coccoliths, foraminifera and ostracods have also been recorded (Warrington and Owens, 1977).

Above the 'Upper Lias', the 'Fuller's Earth Clay' was described by Evans and Thompson (1979) as pale grey, usually silty, shales and mudstones, locally with brown-grey horizons. About 300 m are probably present within the district. Assemblages of common, though restricted, foraminifera and ostracods, with bivalve fragments, small gastropods and fish skeletal remains were reported, but many samples contained no identifiable fauna. The foraminiferal assemblages are similar to those described from the Fuller's Earth Clay (i.e. Bathonian) by Cifelli (1959). A sample from a depth of 8 m in Borehole 72/63 yielded an assemblage of foraminifera and ostracods indicating a probable late Toarcian–early Aalenian age, and suggesting that the borehole was sited near the 'Upper Lias'–'Fuller's Earth Clay' boundary. Palynomorphs (Riding, 1994) and calcareous microfaunas (Wilkinson, 1994) recovered from four gravity core sites (274, 275, 437 and 448 on Figure 29) indicate ages ranging from earliest Aalenian to late Bajocian. According to Evans and Thompson (1979), the overlying 'Oxford Clay' (Callovian to Oxfordian) consists of mottled brown-grey mudstones and clays, with grey shales and silty horizons, and common pyrite. The lower boundary with the 'Fuller's Earth Clay' was 'poorly defined', but a thickness of about 310 m is probably present. The foraminiferal assemblages are comparable to those recorded by Cordey (unpublished PhD thesis, 1963), and common ostracods, bivalves and gastropods are also present.

## Corallian Group

The presence of a small occurrence of Corallian Group in the core of the Bristol Channel Syncline at the extreme west of the sheet is inferred from data in the adjacent sheet area. Its thickness is very uncertain, but may be up to 100 m in the district. These beds are part of the 520 m-thick 'Corallian Sands' of Evans and Thompson (1979). No samples are known from within the district but, to the west, they are described as pale grey or green-grey, clayey, fine-grained sandstone and siltstone, with some thin grey and orange-brown silty shale; glauconite and carbonaceous fragments are locally abundant, but most of the unit is unfossiliferous or very poorly fossiliferous (Evans and Thompson, 1979).

# SEVEN

# Structure

This chapter gives an account of the structure of the rocks which crop out in the onshore part of the district; the deeper concealed structure is described in Chapter 3.

The strongly deformed Devonian rocks contrast with the relatively undeformed Mesozoic rocks present in half-graben basins around Porlock and Minehead. Between Blue Anchor and Watchet, the Late Triassic and early Jurassic rocks seen along the southern margin of the Bristol Channel Basin (Chapter 3) are more intensively deformed and are cut by east–west-trending extensional faults which were partially inverted during the Cainozoic. Faults with a north-westerly trend are numerous in the district but few can be mapped away from the coast; some have a horizontal component. The most noteworthy is the Watchet Fault which can be linked inland with the major Cothelstone Fault that forms the western boundary of the Quantock Hills.

## VARISCAN STRUCTURES

The Devonian rocks were deformed during the Variscan Orogeny, a period of major earth movements which may have begun in the Late Devonian and which continued until Late Carboniferous (Stephanian) times. Various authors (e.g. Simpson, 1969, 1971; Sanderson and Dearman, 1973) have divided the south-west England peninsula into tectonic zones based on fold style. North Devon and west Somerset lie within Zone 1 of Sanderson and Dearman (1973), in which the folds, with axes which are subhorizontal and trend east–west, are overturned to the north and face upwards in this direction. They noted that the amount of overturning varied within the zone. K-Ar dates on slates (Dodson and Rex, 1971) give ages of 330 to 290 Ma for Zone 1. These dates are younger than those for tectonic zones in south Devon, suggesting that the uplift of areas associated with the Variscan deformation occurred at progressively later dates towards the north (Sanderson and Dearman, 1973).

The north–south Variscan compressive forces folded the Hangman Sandstone on approximately east–west to east-south-east–west-north-west axes. The principal fold is the Lynton Anticline, which extends eastwards from the coast east of Lynton (in the Ilfracombe district, Edmonds et al., 1985). In the district its axis is traceable from near County Gate eastwards through West Porlock to Aller-ford; east of there, its course is uncertain (Figure 34). The Lynton Formation occupies the core of the anticline at the western edge of the district, around Malmsmead and Oare, and is faulted on the north against Hangman Sandstone along the Lynmouth–East Lyn Fault (see below). A syncline on the north side of North Hill (Figure 34) separates this structure from a less well-defined anti-cline which probably extends from Bossington Hill to Eastern Brockholes. Between Hopcott and Dunster, northerly dips suggest the presence of a syncline trending through Minehead (Figure 34), but the evidence is uncertain owing to poor exposure.

Folds in the Hangman Sandstone are visible mainly in coastal exposures of the district, and few are seen inland. Fold axes are generally subhorizontal, or show gentle easterly and westerly plunges. Axial surfaces of the folds are inclined to the south at between 20 and 65° (average c.40°). The hinge zones are commonly faulted. The folds are generally open. Locally, monoclinal folds are present, as at Culver Cliff [SS 962 478], near Minehead and near Glenthorne [SS 803 495]. Structures in the Hangman Sandstone at Hurlstone Point, Greenaleigh and Culver Cliff are illustrated in Plates 24–26, and folds near Glen-thorne are shown in Plate 15.

Interbedded argillaceous units within the Hangman Sandstone commonly show axial planar cleavage. Fracture cleavage is locally developed in sandstones in the hinge zones of some folds.

Apart from the Lynmouth–East Lyn Fault, no major faults can be traced inland through the Hangman Sand-stone outcrop, owing mainly to the lack of lithological contrast in the formation. On the coast, small faults with a dominant trend of between north-west and north-north-west are not uncommon in this formation. Shearman (1967) considered that the north-west-trending faults in north Devon and west Somerset are oblique-slip faults, each with a dextral transcurrent component and a down-throw component on its south-west side. Displacements are not always evident, but many of the faults have verti-cal and horizontal displacements of a few metres. The consensus is that many of the faults were Variscan in ori-gin, but were reactivated during the Cainozoic (Shearman, 1967; Edmonds et al., 1985).

The Lynmouth–East Lyn Fault affects the crestal region of the Lynton Anticline, and brings Hangman Sandstone against Lynton Formation; it is a steeply inclined reverse fault. Edmonds et al. (1985) estimated the throw of this fault to be about 1500 m at the western edge of the dis-trict. Ussher (1889) traced the fault through the south-western part of the district, to Brockwell, near Wootton Courtenay. However, this survey found no sound evidence for extending the fault eastwards from Oare, though in view of its magnitude, it is very likely that it does so. An extrapolation south-eastwards along the known trend of the fault indicates a possible extension through Weirwood Common [SS 834 462] and Porlock Common [SS 845 457] to the Nutscale Water south of Lucott Farm [SS 867 449]. The alignment of the Nutscale Water east of the last point may be influenced by the structure, but there is no evidence of the fault in exposures in the stream.

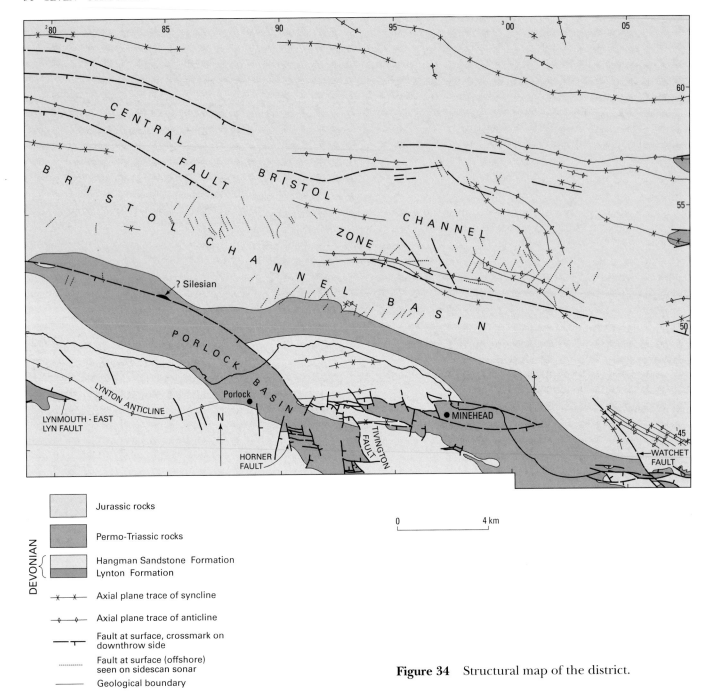

**Figure 34** Structural map of the district.

Edmonds et al. (1985, p.63) noted that there was evidence that the Lynmouth–East Lyn Fault was active locally in Early Devonian times since clasts of older Lynton Formation appear in younger Lynton Formation, the source area lying close to the north. The same authors also considered that the change in structural style from south to north across the east-south-east-trending fault line may reflect a different sub-Hangman Sandstone floor on either side of the fault, which may in turn have contributed to the Exmoor gravity gradient.

Details of structures in the Devonian rocks are given in Edwards (1996).

## POST-VARISCAN STRUCTURES

### Structure of the onshore Mesozoic basins

The Permo-Triassic rocks between Porlock and Tivington and around Minehead occupy half-graben structures in which the major faults occur along the northern margin of the basins. In the Porlock Basin, the Mesozoic sequence extends upwards to include the Penarth Group and part of the Jurassic Blue Lias Formation, but around Minehead, strata above the red Mercia Mudstone are not present.

The main boundary fault along the north side of the Porlock Basin extends from East Lynch westwards through Selworthy to the coast near Bossington; from there it extends offshore to connect with a fault at the southern end of seismic profiles B and C (Figure 11). Onshore, it is locally offset by several small north–south faults. The Mesozoic sequence in the basin dips consistently north. The dips in the Luccombe Breccia probably include an original dip component arising from its deposition on an alluvial fan. Dips in the higher part of the Mercia Mudstone, and in the Penarth Group and Lias, average about 15°, suggesting that this is the approximate amount of northerly tilt which has been imposed on the sequence. Cross sections based on surface dips suggest that the throw on the northern boundary fault is about 1300 m, but the results of gravity surveys (p.28) suggest that the thickness of Mesozoic strata in the basin is much less, indicating a much smaller throw for the main fault. However, interpretation of the gravity data is hampered by the lack of density contrast between the Luccombe Breccia and the bulk of the Hangman Sandstone. The Porlock Basin is bounded on its eastern side by a major north–south fault (the Tivington Fault) which passes through Tivington and is offset by an east–west strike fault through Wootton Knowle. Another important north–south fault (the Horner Fault) occurs east of Horner (Figure 34). Although the intervening ground has not been mapped, a southward projection of the Tivington Fault suggests a connection with the Timberscombe Fault System of Webby (1965), which consists of two subparallel faults averaging about 0.5 km apart. The displacement on these faults has both dextral transcurrent and vertical components; the vertical downthrow across the system was estimated to be 427 m and the amount of dextral displacement about 1.4 km (Webby, 1965). Geophysical evidence (p.24; Figure 8) indicates that a lineament associated with the Timberscombe faults extends northwards to link with the Tivington Fault.

In the western part of the Minehead Basin, the northern boundary fault extends from Hindon Farm [SS 933 467] in the west to Higher Town in the east, and is offset by smaller north–south faults. Between Hindon Farm and Bratton, the southern boundary of the basin is formed by another east–west fault which throws down to the north. South of this fault, a parallel fault extends between Headon Cross [SS 936 459] and Periton and throws down to the south. These two faults bound an uplifted block of Hangman Sandstone which forms a ridge between Little Headon Plantation [SS 943 462] and Little Hill Plantation [SS 948 459], and terminate against a north–south fault through Periton Cross [SS 955 457]. East of the Periton Cross Fault, the form of the Minehead Basin is analogous to that of the Porlock Basin, with the southern boundary being unconformable on Devonian rocks, and dips within the basin being north towards the faulted northern margin at an average of about 15°.

Thomas (1940) considered that the major faults affecting the Mesozoic rocks were initiated in Triassic times and were responsible for the formation of separate basins of deposition. The difference in the Permo-Triassic successions either side of the Horner Fault indicates that

**Plate 24** Syncline in Hangman Sandstone at Hurlstone Point [SS 8991 4916] (GS503).

it was active during deposition of those rocks. The latest movement on faults in the Porlock Basin was post-Liassic.

## Structure of the Blue Anchor–Watchet area

Between Blue Anchor and Watchet, Late Triassic and Early Jurassic strata are extensively faulted and folded. The stratigraphy and structures of the well-exposed foreshore are shown in Figure 35. Cross-sections through the sequence are based on Dart et al. (1995, fig. 5b). These authors note that a horst of Mercia Mudstone, bounded by east–west faults with net extensional displacement to north and south, extends along the foreshore west of Watchet (Figure 35). The horst broadens westwards, towards Blue Anchor, and the structure changes from an anticline to a syncline, cut by additional faults. Well-developed buttress anticlines, which verge towards the horst and locally have overturned forelimbs, are developed in the hanging walls of the faults bounding the northern and southern margins of the central horst. Dart et al. (1995) noted that the southern buttress folding was the

**Plate 25**   Anticline in Hangman Sandstone at Greenaleigh [SS 9511 4818] (GS504). Photograph courtesy of Mr H Prudden.

**Plate 26**   Folds in Hangman Sandstone at Culver Cliff [SS 9614 4783] (GS505).

most complex, with parasitic anticlines and synclines developed around the crest, and a zone of intense folding immediately adjacent to the main fault (Figure 35, Warren Bay sections). Many minor faults cut the buttress zones, and some are at high angles to the main fault.

Dart et al. (1995) and Nemčok et al. (1995) noted that there was evidence from studies of structures in Late Triassic and Early Jurassic rocks at the southern and northern margins of the Bristol Channel Basin of an east–west extensional fault system which developed in response to north–south stretching during Permian to Lower Cretaceous times. This phase was followed by north–south contraction, resulting in partial inversion of the existing faults during Lower Cretaceous to Cainozoic times (Figure 35). Strike-slip faults, such as the Watchet Fault, formed as a result of continuing north–south compression during the Cainozoic.

The Watchet Fault is an important post-Liassic transcurrent reverse fault which crosses the coastline about 1 km west of Watchet (Figure 35); it has been described in detail by Whittaker (1972). The fault is traceable over the foreshore reefs for about 800 m seawards, and forms a distinctive feature in the cliff. On the western side of the fault in the cliff, interbedded reddish brown and green mudstones (Mercia Mudstone) are overlain by Blue Anchor Formation; on the eastern (downthrow) side is a locally disturbed, inverted sequence of Penarth Group and Blue Lias. The Westbury Formation forms a fault gouge 1.5 to 2.5 m wide; the fault plane contact between the fault gouge and the Mercia Mudstone is very sharp and dips south-west at 55°. The main fault surface is not slickensided owing to the softness of the Westbury Formation, but diagonal slickensides are present within the adjacent Mercia Mudstone, and these indicate some horizontal displacement (Whittaker, 1972, pl.XI).

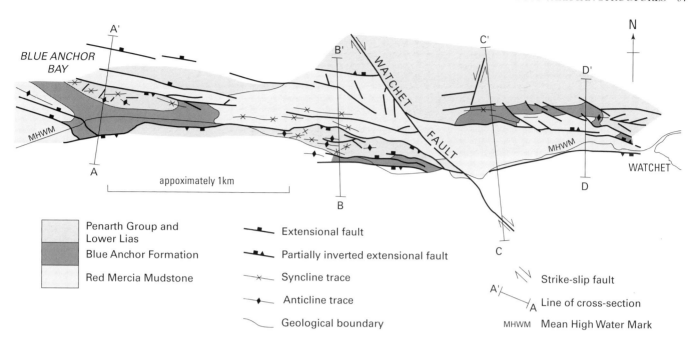

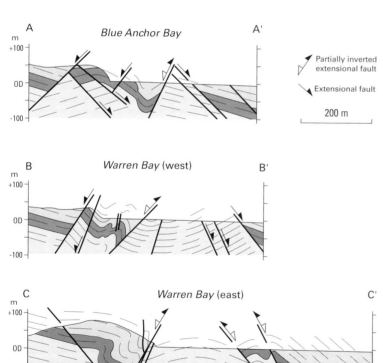

**Figure 35** Geological map and cross-sections of Late Triassic and Early Jurassic rocks on the foreshore between Blue Anchor and Watchet, after Dart et al. (1995). Based on air photographs uncorrected for distortion; scale therefore approximate.

A total vertical displacement in the cliff face of 55 m is indicated by the juxtaposition of *planorbis* Zone strata against red and green mudstones ('Variegated Marl'–Mercia Mudstone).

The transcurrent nature of the Watchet Fault can be shown on the foreshore, where the surface trace forms a prominent line trending at 320°. The fault truncates west- and west-north-west-trending faults and causes dextral drag of their surface traces. The wrench movements have also had the effect of causing the outcrops of east–west-striking beds to curve strongly in towards the fault. Matching of beds belonging to different zones in the Lower Lias indicates a horizontal dextral shift of about 275 m (Whittaker, 1972, p.77).

Dart et al. (1995) noted a north-north-east-trending sinistral strike-slip fault, with an antithetic relationship to the Watchet Fault, on the outer foreshore east of Warren Bay (Figure 35).

The Watchet Fault forms part of the Watchet–Cothelstone–Hatch system of faults, which is traceable for at least 34 km, and is part of a regional pattern of north-west-trending transcurrent faults present throughout much of south-west England. The Watchet Fault probably links with the Cothelstone Fault which partly controls the north-west–south-east alignment of the Quantock Hills. From the southern end of the Quantocks there is a probable south-eastward continuation to link with the Hatch Fault which offsets the Penarth Group escarpment by 2.5 km to the south-east of Taunton.

The main structural features of the Blue Anchor to Watchet coastal section are summarised below. The structures are shown on Figure 35, together with cross-sections through selected parts of the sequence, after Dart et al. (1995).

In the vicinity of Blue Anchor, the most prominent fault extends along the foreshore from a locality [ST 0352 4400] about 500 m north-west of Blue Anchor Cliff to [ST 0474 4377], about 800 m east-north-east of Blue Anchor Cliff (Figure 31). The throw is to the south, bringing an uncertain horizon in red Mercia Mudstone on the north against Blue Anchor Formation, Penarth Group, and Blue Lias Formation (*planorbis* and *liasicus* zones) on the south; a minimum throw of 180 m is indicated. The formations on the south of the fault form a syncline with faulted Lias in the core.

Farther south, about 80 m north-east of Blue Anchor Cliff, gypsiferous Blue Anchor Formation is folded into a shallow anticline (trending at about 100°) in the hanging wall of an extensional listric fault which dips at 32° to 023° (Plate 22). Davison (1994) noted that the fold had probably been tightened during inversion, and small conjugate extensional faults are developed preferentially in the crest of the anticline where layer-parallel extension occurred during folding. The conjugate gypsum veins which are a conspicuous feature of the cliff face were studied by Bradshaw and Hamilton (1967). Three forms of gypsum were observed: 1. pink or white nodules of coarse-grained alabaster, occurring mainly in the laminated mudstones and at the base of massive beds, and probably formed contemporaneously with deposition of the sediments. 2. thin veins of white, finely fibrous satin

spar approximately parallel to the bedding, and folded with it. 3. conjugate veins of pink satin spar which cut 1 and 2.

About 50 m west of Blue Anchor Cliff, a large extensional fault [ST 0380 4366] with a throw of about 80 m brings red Mercia Mudstone on the west against grey Blue Anchor Formation on the east (Front cover). Frictional drag is most marked in the footwall, and is produced by many small synthetic faults which decrease in spacing and increase their displacement towards the main fault (Davison, 1994).

On the foreshore about 500 m east-north-east of Blue Anchor Cliff, Davison (1994) noted a locality [about ST 043 439] on the red Mercia Mudstone outcrop showing normal faults where both lateral tip points can be seen. A footwall anticline and a hanging-wall syncline are present, features which are geometrically necessary for isolated faults. The footwall uplift is not an isostatic effect, but is caused by bending produced by the variable strain along the fault plane. Subsidiary faulting occurs near some fault tips.

About 450 m due north of Warren Farm [ST 051 432], the faulted outcrop of Blue Anchor Formation (Figure 31) lies in the core of a syncline on the foreshore [ST 051 477]. In the cliffs [ST 0536 4343] to the south, steeply dipping Blue Anchor Formation is faulted against Blue Lias (Plate 23).

To the east of the Watchet Fault, a large inverted normal fault [ST 0613 4337] dips at 60° to 167°, but Davison (1994) noted that, at the top of the cliff farther east, the fault dips gently south, suggesting that the fault surface may have been folded during inversion. In the hanging wall of the fault, Davison (1994) noted buttressing of hanging wall strata against those in the footwall. East–west-trending fold axes in the gypsiferous Blue Anchor Formation, parallel to the major fault plane, indicate that approximately north-south shortening took place. Most of the gypsum veins are involved in the folding, but some cross-cut the folds and are post-folding. The Watchet Fault displaces the inverted normal fault by about 500 m with dextral strike-slip movement, and at least 80 m of reverse movement.

Between the Watchet Fault [ST 0607 4332] and West Street Beach, Watchet [ST 0683 4349], a major fault trending between east and east-north-east is present in the cliff, bringing red Mercia Mudstone in the lower cliff against Lias (*angulata* Zone?) in the upper cliff (Figure 31). The fault plane between the Lias and the red Mercia Mudstone in the cliff above is estimated to dip at about 60° south. Eastwards [ST 0635 4342], the fault plane between the Lias and the gypsiferous Mercia Mudstone in the lower cliff is estimated to dip south at about 20°; farther east [ST 0650 4344], the fault plane is estimated to dip south at about 40°. The fault extends eastwards through Watchet Harbour to link with the Doniford Bay Fault of Whittaker and Green (1983, p.101). The fault is well displayed in the cliff [ST 0780 4337] about 200 m east of the district boundary, where it has a throw of about 210 m.

Structures in the Mercia Mudstone west of Watchet were noted by Davison (1994). About 200 m west of West

Street Beach, Watchet [ST 0670 4351], gypsiferous reddish brown mudstones with a few thin greyish green bands are present in the lower part of the cliff. Many small faults show the growth of gypsum fibres along fault planes, indicating that high pore-fluid pressures were operative during faulting, enabling hydraulic separation of fault surfaces and growth of vertical gypsum fibres. The presence of veins with vertical gypsum fibres along bedding surfaces indicates periods during which pore fluid pressure exceeded lithostatic pressure.

Between the Watchet Fault and Watchet Harbour, Blue Anchor Formation is exposed in the core of an east–west-trending pericline [ST 063 437] cut by many minor faults (Figure 31).

# EIGHT

# Quaternary

## INTRODUCTION

The Quaternary (Pleistocene and Holocene) deposits of the district, collectively referred to as 'drift', were formed in a wide range of environments and show corresponding lithological variety. Their distribution is shown in Figure 36. Offshore, the marine deposits comprise gravel, sand and mud formed on the sea bed and in the intertidal zone; onshore, they include saltmarsh and storm beach deposits. 'Submarine forest beds' are associated with the intertidal deposits in Porlock Bay and offshore from Minehead. Fluvial deposits include alluvium (the deposits of modern river floodplains), and river terrace deposits which are the dissected remnants of former alluvium at levels higher than the modern floodplain. Aeolian deposits are represented by small areas of blown sand between Warren Point and Dunster Beach east of Minehead; organic deposits are of small extent, and are represented by peat at one locality in the south-west of the district. Deposits formed by mass movement processes — dominantly head and coastal landslips, with very minor areas of scree and debris cone deposits — are widespread. Made ground, and worked ground and made ground, comprise artificial material deposited by man; the main area of made ground occurs on reclaimed saltmarsh near Minehead.

Large climatic variations characterised the British Isles during the Quaternary, and glacial and periglacial phases alternated with temperate intervals (interglacials and interstadials). Much of Britain was covered by ice sheets during the Anglian, Wolstonian, and Devensian stages, but the district lay south of the limits of the ice sheets, and no glacial or interglacial deposits are known there.

Dating of the Quaternary deposits of the district is hindered by their fragmentary nature, and the lack (except in the Flandrian deposits) of fossils or other material that can be dated by radiometric methods. Consequently, it is not yet possible to produce a satisfactory chronology for the deposits of the district.

### Early to middle Pleistocene

Little is known of the early to middle Pleistocene history of the district. The Bristol Channel was in existence at this time, and the proto-Severn river flowed along it, to open sea to the west. The present-day sea bed slopes to a depth of about 25 m below OD within a kilometre or two of the north Somerset coast, and extends as a gently undulating surface at this approximate depth across the map area to the north. The valley of the proto-Severn is incised into the northern part of the offshore area, trending in an east–west direction with a major kink in the valley at about [SS 930 590] (Figure 36). The cause of the kink is not known: there are no obvious lithological

controls, and it may be caused by a structure which has yet to be identified. The valley has a minimum depth of about 35 m below OD in the east and about 40 m below OD in the west of the district. However, there are enclosed hollows, up to a few kilometres long, within the valley where the thalweg is up to 10 m deeper than the general level. Minor tributary valleys extend into the main channel. Local deepening of the thalweg occurs in the upper part of the Severn estuary where the river and estuarine flow is constricted, but the hollows are smaller than those described here. One possibility is that the hollows are associated with glacial action, though Donovan and Stride (1961) suggested that the rock floor of the valley may have been eroded by the action of tidal sand streams operating during Holocene times.

The age of planation of the -25 m surface and the incision of the main valley is uncertain in the absence of chronological data to constrain the period of erosion. The surface is probably associated with the development of the onshore alluvial terraces of the River Severn; it has not been possible to date these, and the ideas contained in Battiau-Quaney (1984) remain the most considered view on the development of the landscape in the region. The most likely possibility is that the -25 m planation surface was cut by marine action during the late Neogene/early to middle Pleistocene. The planation surface rises gently from west to east into the inner Bristol Channel, and this may be related to the base level to which the process operated — that is, in shallower water, with less wave energy and a shallower wave base platform — or it may signify slight regional warping after incision of the platform.

### Middle to Late Pleistocene glacial/interglacial periods

It is probable that the offshore topography was close to its modern form prior to the late Pleistocene glacial stages, during which sea level fell to expose the offshore area, and downcutting proceeded in the proto-Severn valley.

The district lay at the southern limits of the ice sheets that formed during the Wolstonian and late Devensian glaciations. Patches of ice-transported material (till) on the north coast of the south-west England peninsula, for example at Fremington near Barnstaple and Trebetherick Point in north Cornwall, suggest that the Wolstonian ice sheet impinged against these coasts, and its southward advance may have been similarly halted by the steep cliffs of the district (Kidson, 1977). The southern limit of the most recent (Devensian) ice sheet, which reach maximum extent about 18 000 years ago, lay across South Wales, encroaching seaward to the south into Swansea Bay and Cardiff Bay, some tens of kilometres north of the district.

Although no glacial deposits are known from the district, the proximity of the ice sheets meant that for

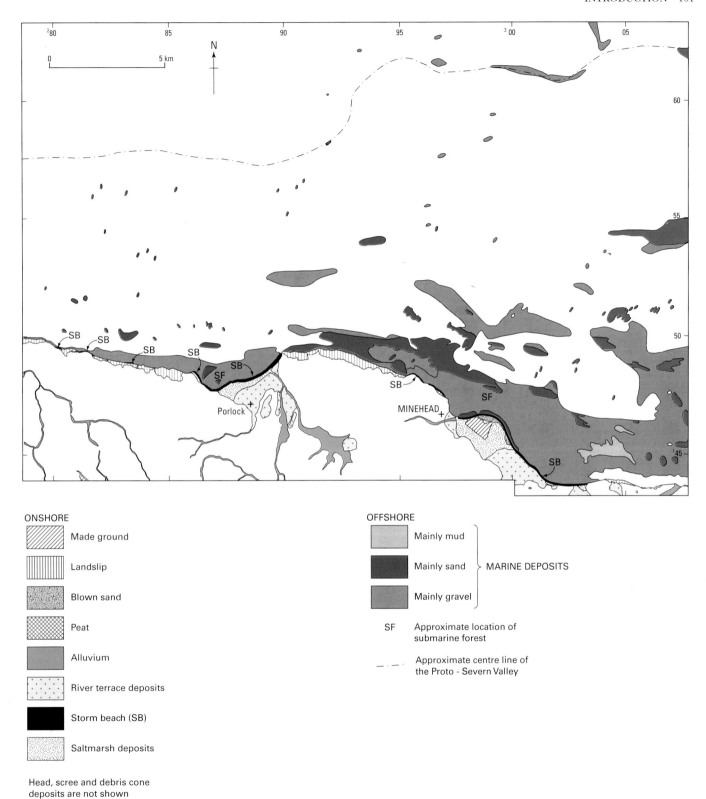

ONSHORE

- Made ground
- Landslip
- Blown sand
- Peat
- Alluvium
- River terrace deposits
- Storm beach (SB)
- Saltmarsh deposits

Head, scree and debris cone
deposits are not shown

OFFSHORE

- Mainly mud ⎫
- Mainly sand ⎬ MARINE DEPOSITS
- Mainly gravel ⎭

SF   Approximate location of
       submarine forest

–·–·–   Approximate centre line of
           the Proto - Severn Valley

**Figure 36**   Distribution of Quaternary deposits in the district.

considerable periods it lay within the periglacial zone. Features formed during earlier glacial stages were, however, probably mostly obliterated during the latest (Devensian) glacial stage (Cullingford, 1982). There was probably widespread formation of solifluction deposits (head) in the district during this period.

Sea level at the maximum of the Devensian glacial stage fell to a depth in excess of 100 m below OD to expose all the offshore area of the district. The proto-Severn valley probably carried much water during each summer melting of the ice sheet and probably even greater volumes during the post-glacial melting of the ice sheet. The degree of down-cutting related to this event is uncertain. Offshore from Barry, in South Wales, north-east of the district, the proto-Severn valley is partially infilled with up to 15 m of material thought to be stiff clay; if this material is of glacial age, then the downcutting of the valley floor in post-glacial time has been minimal.

## Holocene

Most of the offshore part of the district was land during the Devensian glacial phase and until about 10 000 years

before present (BP). Following melting of the Devensian ice sheets, there was a rise in sea level which flooded low-lying coastal areas. Relative sea level (RSL) rose from about 35 m below Mean High Water Spring Tides (MHWST) datum (equivalent to about 30 m below OD) at about 9000 years BP, to about 8 m below MHWST by about 6000 years BP; after this date there was a marked slowing of the rate of rise. This 'Flandrian Transgression' has been well documented from south-west Britain (Hawkins, 1971; Kidson and Heyworth, 1976; Heyworth and Kidson, 1982). Nine new radiocarbon-dated sea-level index points from the 'submarine forest' deposits in the Porlock area, covering the period from about 7800 years BP to about 5000 years BP, confirm the RSL trend recognised by Heyworth and Kidson (1982, fig. 3) (written communication, Dr S C Jennings, March 1996). Pollen analysis indicates that the index points, represented by the vertical contacts between organic and non-organic deposits, are associated with their contemporary MHWST levels. All but one of the new index points lie within an 'error envelope' based on the RSL rise data in Heyworth and Kidson (1982) (Figure 37). The outlying point (7730 ± 50 years BP) suggests that the rate of rise between 8000

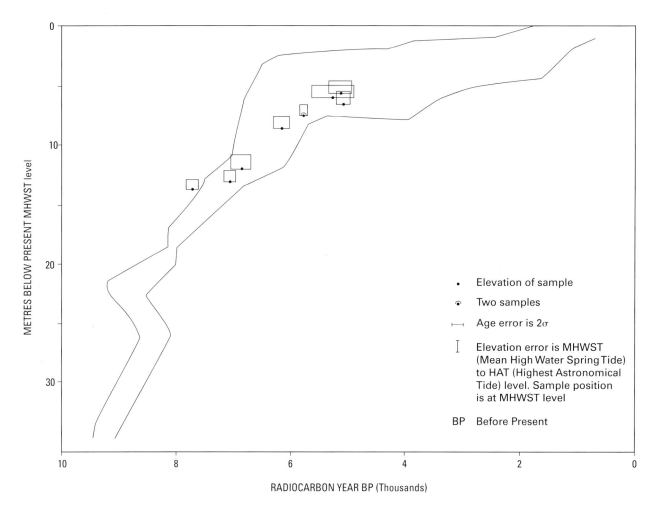

**Figure 37** Graph of age and altitude of radiocarbon-dated samples from the Porlock Bay area (Dr S C Jennings, written communication, March 1996).

and 7000 years BP was slower than that suggested by Heyworth and Kidson's data; however, there are very few index points of this age and elevation on Heyworth and Kidson's graph, and the early Holocene RSL trend in the Bristol Channel is thus poorly constrained (written communication, Dr S C Jennings, March 1996).

## MARINE AND COASTAL DEPOSITS

### Saltmarsh deposits

The main areas of saltmarsh deposits in the district form low-lying ground behind the Porlock Bay storm beach gravel ridge (see below), and east of Minehead. Minor areas are present behind the shingle ridge at Greenaleigh, probably concealed beneath made ground north-west of Minehead Harbour; west of Blue Anchor Station; and at the mouth of the Pill River, Blue Anchor (Figure 36).

The inactive saltmarsh deposits of Porlock Bay are periodically inundated when the shingle ridge is breached, an event which occurred in 1981 and 1989/90 and, most recently, at the end of October 1996. During the winter of 1989/90, the area was flooded to a depth of 0.6 to 1.2 m, and remained submerged in salt water for several weeks.

Godwin-Austen (1865) reported that the deposits to landward of the shingle ridge consist mainly of 'marine silt'. The uppermost unit consists of yellowish brown mud containing shells of Scrobicularia piperata with the valves united. Beneath this 'Scrobicularian mud' Godwin-Austen noted a 'band of vegetable matter' overlying dark tenacious clay deposits.

The stratigraphy of the deposits proved in boreholes beneath Porlock Marsh on the landward side of the shingle ridge is shown in Figure 38 (Dr S C Jennings, written communication, March 1996). The deposits are mostly younger than those on the seaward side of the ridge (see under 'submarine forest', below), on the evidence of radiocarbon dates of 5250 ± 180 and 5140 ± 100 years BP from a peat near the base of the proved sequence. Overlying the peat are about 4 m of mainly non-organic sands and silty clays with marine shells.

The inactive saltmarsh deposits east of Minehead form an extensive coastal flat at about 5 m above OD, partly occupied by industrial estates and a large holiday camp, but with areas of marsh locally present. Trial pits and boreholes show the deposits to be predominantly clay up to a maximum recorded thickness of nearly 12 m. South of Butlin's 'Somerwest World' Holiday Camp, a division of the saltmarsh deposits into an upper unit of brown clay overlying grey clay was recorded in some boreholes (Figure 24). No basal gravel was present in these boreholes, but in the vicinity of Mart Road [SS 9750 4595], near the western limit of the deposit, up to 3 m of clay overlies gravel, the full thickness of which was not proved. The general sequence proved in trial pits along the southern route of Seaward Way is 0.15 m of soil overlying up to 1.7 m of firm, grey-brown, silty clay, on up to 1.9 m of soft, grey-brown, silty clay becoming blue-grey with depth; peaty material was recorded in one trial pit at 2.1 m

depth. Close to their southern boundary near Ellicombe, the saltmarsh deposits rest on head, and in one borehole [SS 9865 4476] 2.9 m of blue-grey and brown mottled clay (saltmarsh deposits) were recorded overlying at least 1.1 m of red sandy clay with sandstone fragments (head).

Further details of trial pits and boreholes penetrating the saltmarsh deposits are given in Edwards (1996).

### 'Submarine forest'

The presence of 'submarine forest' deposits in the intertidal zone in Porlock Bay and offshore from Minehead has long been known, and there is a brief account in De la Beche (1839, p.419). The fullest account is by Godwin-Austen (1865). The deposits are not shown separately on the map, but their approximate extent is indicated by the words 'Submarine Forest' printed on the intertidal zone of the Ordnance Survey base map.

In Porlock Bay, the 'submarine forest' beds are reported to be visible at very low tides in the intertidal zone seaward of the storm beach ridge which runs unbroken along the length of the bay. The associated sediments are thin (deduced from Godwin-Austen (1865) to be less than about a metre) and rest on head. These were not seen during the survey; most of the lower intertidal zone is a flat pavement, made up of well-rounded cobbles with an upper coating of biological growth, giving them a dark grey-brown colour, which suggests that they are not presently being moved by wave action. Landward and under the shingle ridge lies marine silt, which passes down into a surface of plant growth several centimetres thick, a thin freshwater mud, and the remnants of a 'forest' resting on angular head. The stools of the trees rise above the level of the plant growth and upon this level lie the tree trunks, which are in places over 6 m long and did not sink into the surface on which they fell. Under the surface of plant growth is a tenacious blue mud which surrounds the trees but has no roots in it, suggesting that the mud accumulated subsequent to the 'forest' growth. The mud was assumed to be of fresh water origin, from the abundance of diffused vegetable matter. The underlying 'submarine forest' consists of tree stools, up to 0.7 m in diameter. The largest are of oak, distinguished by their black colour; others are red and are probably alder. Flooring the blue mud is angular debris, probably head, which accumulated in the intertidal zone during postglacial time. The trees forming the 'forest' were rooted into this lithology.

Recent work (e.g. Cooper et al., 1995) has added considerably to knowledge of the Holocene development of the Porlock coastal environment since the account of Godwin-Austen. Dr S C Jennings (written communication, March 1996) has contributed the following preliminary results of a programme of trial boreholes and radiocarbon dating in the 'submarine forest' beds of Porlock Bay. Boreholes [centred on SS 871 478] in the 'submarine forest' beds seaward of the shingle ridge showed up to c.5.4 m of silty clays with thin organic horizons, resting on head; detailed sequences penetrated in the boreholes, together with radiocarbon dates, are shown in Figure 38. The sequence represents a series of alternations between

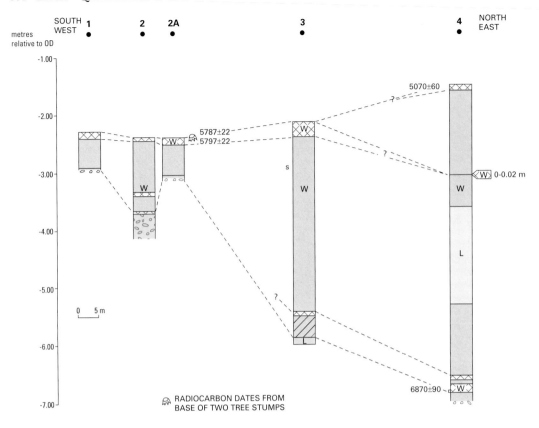

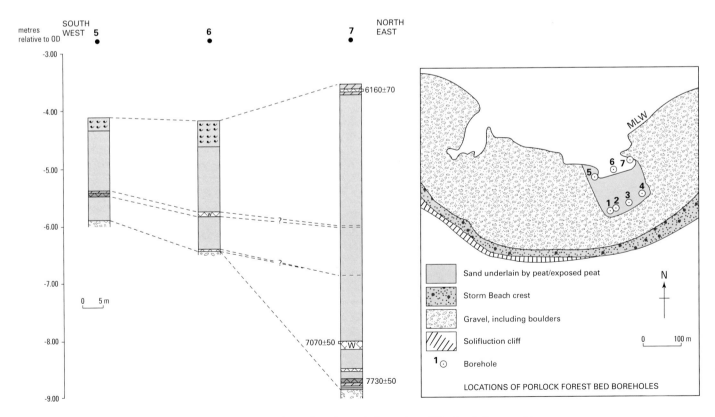

**Figure 38**  Stratigraphy of boreholes in the 'submarine forest' deposits of Porlock Bay and in the deposits beneath Porlock Marsh (Dr S C Jennings, written communication, March 1996).

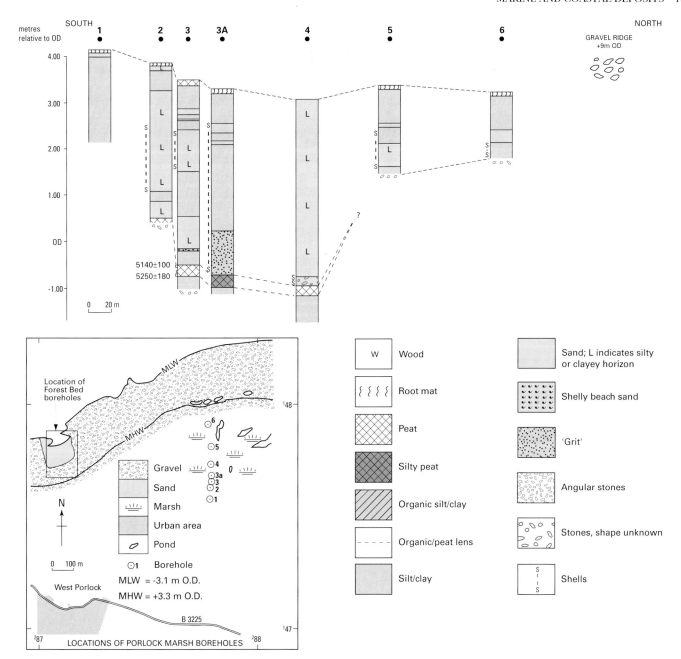

**Figure 38** *continued.*

freshwater (organic) and estuarine (non-organic) conditions between dates of 7730 ± 50 years BP and 5070 ± 60 years BP; sharp boundaries above and below the organic deposit suggest that the environmental changes were the result of opening and closing of tidal inlets within the Porlock Bay barrier beach system. Pollen analysis indicates that alder carr was established during the organic-deposition episodes, while the non-organic deposits formed in low-energy estuarine conditions established by inlets through a barrier beach (Cooper et al., 1995). The location of former tidal inlets may have been governed by the subsurface topography of the head deposits.

Offshore from Minehead, the 'submarine forest' deposits occur low in the intertidal zone at about 2 to 4 m below

OD. Kidson and Heyworth (1976) recorded a peat bed at between 1 m to 3 m below OD off Stolford in Bridgwater Bay, about 15 km east of the Minehead district.

**Storm beach**

Storm beach gravels are developed as narrow ridges along the coast of the district, mainly in Porlock Bay, between Greenaleigh Point and Minehead, and from Minehead to Blue Anchor. Sporadic patches also occur along the coastline west of Porlock Weir, especially at Embelle Wood (Figure 36).

The storm beach gravels are well developed in Porlock Bay, where they form a ridge commonly regarded as one of the finest examples of its type remaining in England (Plate 27). It extends for about 4.5 km between Gore Point and Hurlstone Point. Landward of it lie mainly salt-marsh deposits or, between Porlock Weir and Porlockford, and near Hurlstone Point, head deposits forming low cliffs (p.109). To seaward, most of the lower intertidal zone is a flat pavement made up of well-rounded cobbles with an upper coating of biological growth, giving them a dark grey-brown colour, which suggests that they are not presently being moved by wave action. This pavement passes abruptly landward into the long gravel ridge which is breached only at Porlock Weir. The Horner Water north of Bossington drains through the ridge. The height increases from west to east and the ridge also widens in this direction.

Most of the cobbles forming the ridges are made up of Devonian rocks derived from the cliffs to the west. It is uncertain how much material is being supplied by active cliff erosion, and how much is derived from erosion and landward movement of the offshore head platform. The latter, when exposed in the intertidal zone off Porlock, appears to be stable, though in Dunster Bay the gravel is formed into ridges which may move in time to provide material to the shoreface.

Comparisons between the western (Gore Point) and eastern (Hurlstone Point) ends of the ridge have been made during studies at the Field Studies Council at Nettlecombe Court, Williton (written communication, Mrs H J Wilson, February 1996). These show that material at the eastern end of the ridge is better sorted, smaller and rounder than that at the western end (Figure 39). At Hurlstone Point most of the clasts are very well rounded and between 5 and 10 cm maximum pebble length.

The form of the ridge changes along its length, due to both man-made and natural processes. In the west, the ridge is simple in form, with a well-defined landward and seaward slope and a sharp crest. This form is best developed north of Porlockford. To the east, the crest of the ridge becomes flatter, cusps of gravel are attached to the landward edge, and a series of scarps related to storm levels occur on its seaward slope. In the east, the pebbles on the crest of the ridge are locally covered in lichen and the ridge appears to be stable. Groynes have been constructed along the western part of the ridge in order to limit the west–east transport of the cobbles. The effects of these groynes are most noticeable at Porlock Weir, where piles along the western river mouth show evidence of being pushed over by the mass of cobbles many metres

**Plate 27** Storm beach shingle ridge at Porlock Bay, looking towards Hurlstone Point at the northern end of the ridge (September 1992) (GS506).

high. The effect of these groynes at Porlock Weir has been to starve the ridge to the east. Although additional groynes have been constructed north-east of Porlockford, they have been overtopped. The ridge is formed into two major cusps joined at about [SS 877 480] where a drain, from the meadows behind, extends onto the intertidal zone; this is the site of most of the present (1995) erosion and instability on the ridge.

West of Porlock Weir, a small storm beach ridge is developed at Glenthorne Beach [SS 801 495], but a more continuous shingle ridge is developed for about 1.5 km from east of Glenthorne to Broomstreet Combe [SS 8052 4944 to 8179 4916]. North of Embelle Wood, the ridge lies to seaward of a head-covered slope, and includes sandstone boulders up to 0.4 m in diameter. Farther east, a gravel ridge about 200 m long is developed on the coast [SS 8335 4883] north of Culbone Wood.

East of Hurlstone Point, small patches of storm beach gravel are present at three places along the coast [SS 9215 4926; 9247 4923; 9335 4886]. A more continuous ridge flanks the head and saltmarsh deposits at Greenaleigh and extends to Minehead Harbour, broken only at Culver Cliff. The shingle ridge continues from a place [SS 9740 4641] just north of Minehead Station eastwards round Warren Point for about 7 km, to near Blue Anchor [ST 0327 4355].

### Sea-bed sediments and intertidal deposits

Prior to the Flandrian Transgression, the present sea bed was covered by periglacial and glacial outwash debris, and the river courses were floored by alluvial sediment. The incursion of marine water into the inner Bristol Channel resulted in strong tidal currents which removed much of this material from the sea bed, rolling it forward with the advancing Holocene shoreline. The distinctive dip and scarp microtopography of the bedrock exposed at the sea bed suggests that it is being slowly etched by sand-loaded currents.

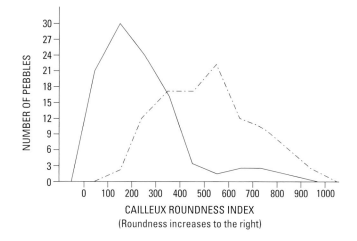

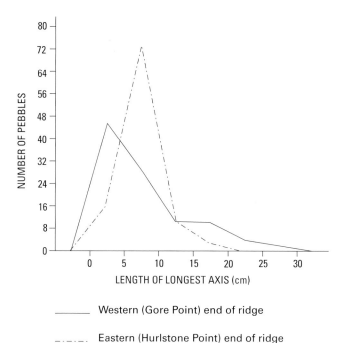

_____ Western (Gore Point) end of ridge

_._._. Eastern (Hurlstone Point) end of ridge

**Figure 39** Comparison of the size and roundness of clasts at the western (Gore Point) and eastern (Hurlstone Point) ends of the Porlock Bay shingle ridge. Data from the Field Studies Council at Nettlecombe Court, Williton.

The distribution of sea-bed sediments in the offshore part of the district is shown on Figure 36. Seismic and sidescan sonar data show that across most of the offshore area, the Holocene sediments are very thin (less than a metre thick) or absent, and bedrock is at or close to the sea bed. A narrow tongue of sand [SS 070 540] is the western limit of Culver Sand, a major sand bank aligned approximately parallel to the main tidal flow. The bank rests on a flat bedrock platform and its crest (east of the district) breaks surface at low water. Elsewhere, small, probably ephemeral, sand patches and ribbons are developed. The proto-Severn valley is locally floored by gravel

[e.g. SS 991 612–ST 033 611], assumed from its seismic character to be of Holocene age.

Thicker accumulations of offshore Holocene sediments occur in a belt parallel to the coast and, in particular, in Blue Anchor Bay. Extensive spreads of gravel cover the sea bed off the coast from Warren Bay to Minehead. To the west, a narrower band of coastal sand and gravel extends to Hurlstone Point, and gravel again is present off Porlock Bay. To the west of the bay, nearshore sediments — mainly gravel — are restricted to a narrow band below the low water mark.

The intertidal zone in Warren Bay is a wave-cut platform with a thin and variable cover of mud, sand and gravel. This passes landward into an eroding steep cliff of Mesozoic strata locally fronted by a narrow, steep, gravel beach. Seaward, a thicker gravel and locally sand cover overlies the bedrock platform. In Blue Anchor Bay to the west, the intertidal zone displays a predominantly gravelly lower section with broad expanses of flat gravel pavement, and locally shows low gravel ridges up to about a metre high and a few hundred metres long. The upper part of the zone is sandier and passes into a continuous shore-parallel storm beach gravel ridge in the high tide zone. The distribution of the gravel around the old and new groynes in this area indicates active transport of the gravel eastwards. The outcrop of early Holocene 'submarine forest' tree-stools and blue-grey clay off Minehead demonstrates that these sand and gravel deposits are generally thin (probably less than a metre) across much of this area and the greater part of the Quaternary sequence is made up of Pleistocene head. The seismic evidence suggests that the head rests on a wave-cut platform which is a continuation of the modern beach rock platform. This implies that the offshore platform is older than the Holocene, and that the present beach platform is an inherited feature that has been modified during the Holocene rise in sea level.

The movement of sediment in the Bristol Channel has been studied for over a century. Sollas (1883) suggested that the mud infilling Bridgwater Bay and the Somerset levels had a seaward source. Culver and Banner (1978) supported this idea, and noted that the mud contained planktonic foraminifera with an oceanic source; however, they concluded, from the orientation of sand waves, that the sand-sized sediment fraction behaved differently to the mud, and was transported down-estuary. An examination of the heavy minerals (Barrie, 1980) suggested that there is eastward movement of sand along the north Somerset coast, a view which is supported by the evidence from the movement of the materials making up the gravel ridges. These studies, and models related to bedform morphologies, showed a complex pattern of offshore sediment transport in the district, with a dominance of westward ebb-dominated transport.

McLaren and Collins (1989) attempted to identify sediment transport pathways in the Severn estuary and Bristol Channel by applying statistical techniques to the grain-size analyses obtained from about 900 samples taken across the region. They reported that the rivers draining into the estuary (especially the River Parrett) were the source of the mud in Bridgwater Bay. They suggested that, in the Blue Anchor Road area, between Culver Sand and the north Somerset coast, sand and mud are both being transported to the west, that the dominant grain size is coarse sand and that net erosion is occurring in the area. They recognised some eastward drift of sand along the coast at Minehead, but did not report on the transport direction of the gravel in the shoreface ridges. The suggestion that the movement of sand is to the west is supported by the asymmetry of the sandwaves, which is shown on the 1:50 000 Series map. McLaren and Collins (1989) also demonstrated local reversal of transport direction in the area south-west of Culver Sand and off Hurlstone Point. The sea-bed sediment in the map area to the west of Blue Anchor Road was so thin or coarse-grained that McLaren and Collins (1989) did not attempt to infer a net sediment transport direction.

The conclusion from these studies is that the offshore sand is moving westwards across the area whilst the material making up the shorebound gravel ridges is moving eastwards.

## FLUVIAL DEPOSITS

### River terrace deposits

River terrace deposits comprise the dissected remnants of former alluvial deposits now occurring at levels above the present river floodplains. In the district, they consist mainly of gravel.

Low-level, gently sloping (less than 2°) spreads of gravel are present landward of the saltmarsh deposits in the Vale of Porlock, between Porlock and Bossington and around Allerford. Similar low-level gravel spreads are also present east of Minehead between Marsh Street and Blue Anchor (Figure 36). The few exposures in the low-level deposits near Porlock indicate that they consist of brown, poorly bedded gravel containing subrounded to subangular Devonian sandstone clasts, mainly less than 10 cm but locally up to 15 cm across, and fairly closely packed in a matrix of brown, silty sand. The deposits east of Minehead comprise gravel with clasts, between 2 mm and 0.35 m diameter (mostly less than 0.15 m), of red, fine- to medium-grained sandstone and shale, with minor tuff, porphyritic spilite and quartz. Indistinct bedding is picked out by the horizontal alignment of the long axes of clasts; a few sand lenses are present. The maximum thickness of the low-level river terrace deposits is uncertain; trial resistivity soundings [SS 8857 4788 and 8869 4736] made north of Porlock indicated a high resistivity layer overlying a lower resistivity layer; interpreting the upper layer as river terrace deposits indicates thicknesses of 22 m and 34 m (bases at both sites at about 17 m below OD).

In the valley of the Chalk Water, south of Oareford, minor terraces are present at two places [SS 8197 4490; 8195 4462] and consist of pebble and boulder gravel with sandstone clasts.

Older river gravels form terraces at higher levels. In Porlock Vale, three well-developed terraces overlie the Mercia Mudstone between Porlock and Holnicote (Figure 36). The largest spread, near Porlock, has an upper surface sloping gently north at about 2° from about 70 m above in the south to 40 m above OD in the north. The upper surface of the most easterly terrace is at about 60 m above OD. The few exposures (e.g. [SS 8981 4742] in river cliffs just west of the Horner Water) show the river terrace deposits to consist of up to 5 m of brown and orange-brown, poorly bedded gravel, with subangular to subrounded and rounded pebbles and cobbles of Devonian sandstone up to about 0.15 m across in a matrix of fine gravelly silt.

Another sloping spread of gravel occurs between Blackford and Tivington; exposures [SS 9273 4532] in Long Lane show 0.5 m of orange-brown, gravelly sand with interbeds of sandy gravel, overlying 0.5 m of gravel with angular clasts of Devonian sandstone, mainly less than 5 cm across.

In the Blue Anchor area, Carhampton Knap [ST 0065 4337] is capped by a small area of river gravel at about 50 m above OD. A more extensive terrace [centred on ST 044 433], about 1 km east of Blue Anchor Hotel, rises to 85 m above OD; there are no exposures, but the soils contain subrounded sandstone, slate and vein quartz clasts. The thickness probably does not exceed 3 m.

Further details of the river terrace deposits of the district are given in Edwards (1996).

### Alluvium

Alluvium, the deposits of modern river floodplains, occurs as narrow strips bordering the larger streams on the Hangman Sandstone outcrop west of Porlock, as more extensive spreads in the Vale of Porlock and, to a minor extent, in Minehead (Figure 36). In the narrower streams on the Devonian outcrop, the alluvium is commonly gravel, locally very coarse and with boulders of Devonian sandstone. The larger streams, such as the Oare Water, show an upper loam unit about 1 m thick resting on gravel. In the Vale of Porlock, the alluvial flat along the Horner Water is ill defined and not easily distinguished from the adjacent slightly higher-level river terrace deposits. The deposits along the Horner Water are locally very coarse grained, and a typical river-bank section [SS 8993 4677] north of New Bridge, West Luccombe, shows 1.5 m of brown gravel with boulders and pebbles of Devonian sandstone.

The alluvium in the Vale of Porlock east of Holnicote consists of a variable sequence from 0.3 to over 1 m thick of silt, clay and sand — locally with organic clays containing wood fragments — which rests on gravel. From south of Holnicote to Blackford, the alluvial flat broadens to about 0.6 km wide where the alluvium overlies the Mercia Mudstone outcrop.

Further details of the alluvium are given in Edwards (1996).

## ORGANIC DEPOSITS

### Peat

Peat has been mapped at only one locality [SS 812 440], near South Common, in the south-west of the district

(Figure 36). It consists of black clayey peat occupying an ill-defined boggy hollow at the head of Stowford Bottom.

## AEOLIAN DEPOSITS

### Blown sand

Deposits of wind-blown sand are present east of Minehead between Warren Point [SS 984 464] and Sea Lane End [ST 005 445], east of Dunster Beach (Figure 36). They lie between the shingle ridge on the seaward side and salt-marsh deposits or river terrace deposits to landward. No good sections are available, but the soils indicate that the deposit is mainly brown-weathering, fine-grained sand. Numerous arcuate ridges evident on air photographs within the blown sand area, and apparently of natural origin, may reflect earlier stages in the development of the storm beach shingle ridge, subsequently blanketed by blown sand.

## MASS-MOVEMENT AND RESIDUAL DEPOSITS

### Head

De la Beche (1839) used the term 'Head' for rubbly slope deposits in coastal sections of south-west England. It has since been widely used to describe structureless or poorly bedded mixtures of clay, silt, sand and stones which are believed to have moved downslope mainly by solifluction (soil-flow) in a periglacial climate. In these conditions, alternate freezing and thawing of the surface layers resulted in a slow downslope flow of water-logged soil and other unsorted material. Soil creep, hill wash, and other mass movement processes are still active in the movement of surface materials. During the deglaciation stage, the high rainfall would have aided movement of regolith down from steep slopes onto and across the inner coastal lowlands, thus explaining the probable presence of head in the nearshore zone between Porlock Bay and Minehead.

Two categories of head have been recognised in the district: blanket head and regolith, and valley head (the latter termed 'Head' on the 1:50 000 Series map).

Deposits east of Watchet, near the eastern margin of the district [ST 074 432], are shown on the map as head; they are contiguous with gravels exposed on the coast of the adjacent district at Doniford, which have yielded remains of woolly mammoth (*Elephas primigenius*) and Palaeolithic implements of Acheulian type (Wedlake, 1950; Wedlake and Wedlake, 1963). The deposits were studied by Gilbertson and Mottershead (1975), who concluded that they were periglacially reworked river gravels attributable largely to the Devensian glacial stage.

Seismic profiles off the north Somerset coast between Minehead and Hurlstone Point show a distinctive flat sea bed underlain by an extensive tabular unit over 5 m thick, with a chaotic seismic signature. There are no samples from the unit, and the seismic evidence suggests that it consists of coarse-grained gravel; the deposit is most likely to be head, although the possibility that it is of glacial origin (till) cannot be wholly excluded.

### BLANKET HEAD AND REGOLITH

The Devonian rocks of the district are nearly everywhere mantled by a deposit composed partly of in situ weathering products of the parent rock ('regolith'), and partly of material that has been subject to mass-movement processes (including solifluction). It is not possible to determine to what extent, if at all, regolith material has been transported by solifluction or other mass-movement processes, and the term 'blanket head and regolith' is used for this material. The deposit is so widespread that it has been omitted from the map for the sake of clarity. The thickness of the blanket head and regolith is typically between 2 and 3 m, but greater thicknesses are probably present locally.

The deposit typically consists of orange-brown to reddish brown, unbedded to poorly bedded gravel, composed of angular clasts of Devonian sandstone in a poorly sorted matrix of gritty, clayey, sandy silt, but it locally reflects the underlying lithologies and may be shale-rich where underlain by shale. At some localities there is a high proportion of angular sandstone in the deposit, giving rise to a rubbly, angular, matrix-poor deposit; this may in places represent fossil scree material (see below). Clasts mostly are smaller than 0.1 m, but larger angular boulders are present locally.

Small sections of blanket head and regolith are widespread in road and track cuttings and in quarries; descriptions of a selection of typical localities are given by Edwards (1996).

### VALLEY HEAD

Valley Head (referred to on the published 1:50 000 Series map as 'Head') typically occurs as strips along valley bottoms and sides. On the Hangman Sandstone, the head forms narrow outcrops along the valley floors; on Permo-Triassic rocks it forms wider outcrops along the valley bottoms and sides, and these coalesce to form more widespread areas of head. For example, in Porlock Vale between Porlock Weir and Porlock, the valley head forms a sloping apron lying between the steep hills of Devonian sandstone to the south and the saltmarsh or river terrace deposits to the north. A similar apron is present in the Minehead area, extending between the centre of the town, south-eastwards through Alcombe to near Marsh Street.

The valley head consists predominantly of poorly sorted mixtures of silt, sand and clay with angular rock fragments, mainly of Devonian sandstone. Near Blue Anchor, head overlying Mercia Mudstone commonly contains fragments of Lias limestones.

In Porlock Bay, the mid-Holocene 'submarine forest' (p.103) is reported (Godwin-Austen, 1865) to rest on head, in which the 'forest' trees were rooted.

The coastal head in the Porlock area is typified by easily accessible exposures in low cliffs [SS 8677 4763 to 8700 4762] at Porlock Beach (Plate 28) which show 3 m of brown to orange-brown gravel with angular to subangular and a few subrounded clasts of grey and purple, fine-grained sandstone up to 0.3 m across, in a matrix of granular, sandy silt. The deposit is unbedded to crudely

**Plate 28**  Head at Porlock Beach [SS 8690 4762], Porlock Weir. The hammer is 0.3 m long (GS507).

bedded, with subparallel alignment of clast long axes. There are some local areas of open-framework pebble and granule gravel. There is little apparent lithological variation along the cliff, but at the eastern end a few impersistent sand beds are present. The modern shingle ridge is banked up against the head cliffs. A Lower Palaeolithic implement of Acheulian type was recorded by Grinsell (1970, p.13) as possibly being derived from 'the lower gravels exposed in the cliffs'; this may refer to the low cliffs of head at Porlock Beach.

Near Minehead, typical exposures of head are present in cliffs [SS 9530 4818] towards the western end of the arcuate ridge of storm beach shingle at Greenaleigh. The head is 2 m thick on the east, thickening to 6 m on the west, and consists of orange-brown, sandy gravel with angular to subrounded clasts of Hangman Sandstone, mostly less than 0.1 m across, up to a maximum observed diameter of 0.3 m. The orientation of flat clasts defines bedding subparallel to the ground surface. In one place [SS 9527 4817] at the base of the cliff, well-rounded pebbles up to 0.2 m across are apparently incorporated in about 0.3 m thickness of the head; these pebbles may indicate the former presence of raised beach deposits in

this area (Prudden and Edwards, 1994), but it is considered more probable that pebbles from the beach shingle have been forced into the head during storms.

Head on the cliff top east of Blue Anchor [ST 0332 4355 to 0380 4366] is largely inaccessible, but apparently consists of about 0.9 m of brown clay on about 1.0 m of brown gravel with Lias limestone clasts.

Further details of coastal and inland exposures of valley head are given by Edwards (1996).

### Debris cone

Small areas of debris cone deposits have been mapped at two localities [SS 8251 4634; 8272 4623] on steep slopes on the south side of the valley of the Weir Water, about 1.5 km east of Oareford. They consist of cone-shaped masses of debris washed down small, actively eroding gulleys, and piled up at the base where the gradient decreases at the margins of the alluvial flat. The transported material is apparently derived mainly from blanket head and regolith, so that the composition of the debris cones is lithologically similar to the head.

### Scree

Scree in the district consists of angular sandstone rubble developed on steep slopes on the Hangman Sandstone outcrop; occurrences of mappable extent are shown on the map. In the adjacent district to the west, scree development was attributed to accumulation in a periglacial climate during the Devensian (Edmonds et al., 1985).

The most extensive development of scree is on about 32°, east- and south-facing slopes at Parsonage Cleave [SS 792 485] on the north-east side of the East Lyn River, about 800 m north of Malmsmead. The scree material is mainly grey and green, angular Hangman Sandstone. The other main areas of scree — also mainly angular Hangman Sandstone rubble — occur on the sides of steep-sided valleys (combes) on the north side of Bossington Hill and North Hill, in particular along the north side of Hurlstone Combe, on the east side of East Combe, on the north-east side of Henner's Combe, and near Furzebury Brake.

At some localities [for example SS 8907 4389], on slopes north of the Horner Water, about 1.5 km south of Horner, blanket head and regolith is locally overlain by a matrix-free deposit of angular, open-framework, sandstone rubble, which may represent fossil scree material now stabilised by soil and vegetation; these occurrences suggest that scree development may formerly have been more widespread in the district.

### Landslip and coastal erosion

Landslips are an important feature of the coastlines of the district, and are widely developed on both the Devonian and Triassic–Jurassic rocks; slips are rare inland. On the Hangman Sandstone, the main areas of former and present instability are on the coastal slope west of Porlock Weir as far as Glenthorne, and between Hurlstone Point and Greenaleigh (Figure 36). West of Porlock Weir, recent slips in Yearnor Wood have affected the South West Coast

Path, and consequently investigations into the cause of the instability have been carried out (Anon., 1992, 1994). Between Blue Anchor and Watchet, the varied rock types and the structural complexity have resulted in a wide range of instabilities, such as block slides, rockfall, mud flows, and rotational failures, affecting the Triassic and Liassic rocks in a relatively short stretch of coastline (Figure 36).

An account of the landslipping is given in the slope stability section of Chapter 2.

COASTAL EROSION

Williams and Davies (1990) estimated that the Devonian cliffs on the south coast of the Bristol Channel are receding at about 4 cm/year, a rate five times slower than that estimated for the coast formed of Triassic strata. Whilst this appears to be slow, it accumulates into recession of the Devonian cliffs of some 200 m over the last 5000 years, and a kilometre of the Triassic cliffs. Using these values suggests that the shorebound depositional features, primarily the gravel ridges, contain up to 200 years input of sediment derived from coastal erosion. If these estimates are correct, the depositional features contain only a fraction of the material eroded over the past 5000 years.

## ARTIFICIAL DEPOSITS

### Made ground, and worked ground and made ground

Made ground consists of artificial material deposited on the original ground surface; worked ground and made ground comprises artificial deposits filling disused pits and quarries. An account of these deposits is given in Chapter 2.

# INFORMATION SOURCES

## BGS PUBLICATIONS DEALING WITH THIS DISTRICT AND ADJOINING DISTRICTS

### Books

*British Regional Geology*
South-West England, 4th edition

*Memoirs*
Ilfracombe and Barnstaple (sheets 277 and 293), 1985.
Weston-super-Mare (sheet 279), (1983).
Taunton and the Quantock Hills (295), (1985).
Wells and springs of Somerset (1928). (Out of print).
The metalliferous mining region of South-West England, Vol. 2 (1956; Addenda and corrigenda 1988; reprinted 1994).

*Special Reports on the Mineral Resources of Great Britain*. Vol. 3. Gypsum and anhydrite, (3rd edition, 1938). (Out of print).
*Special Reports on the Mineral Resources of Great Britain*. Vol. IX. Iron ores (contd.). Sundry unbedded ores of Durham, East Cumberland, North Wales, Derbyshire, the Isle of Man, Bristol District and Somerset, Devon and Cornwall (1919). (Out of print).
*Special Reports on the Mineral Resources of Great Britain*. Vol. XXI. Lead, silver-lead and zinc ores of Cornwall, Devon and Somerset (1921). (Out of print).
*Special Reports on the Mineral Resources of Great Britain*. Vol. XXVII. Copper ores of Cornwall and Devon (1923). (Out of print).

*United Kingdom Offshore Regional Reports*
The geology of Cardigan Bay and the Bristol Channel (1994).

### Maps

*1:584 000*
Tectonic map of Great Britain and Northern Ireland

*1:625 000*
South Sheet (Geological)
South Sheet (Quaternary)
South Sheet (Aeromagnetic)

*1:250 000*
Bristol Channel (Solid geology)
Bristol Channel (Sea bed sediments)
Bristol Channel (Gravity)
Bristol Channel (Aeromagnetic)

*1:100 000*
Hydrogeological map of the Permo-Trias and other minor aquifers of South West England

*1:50 000 and 1:63 360*
Sheet 277 (Ilfracombe) (1981)
Sheet 293 (Barnstaple) (1982)
Sheet 279 (Weston-super-Mare) (1980)
Sheet 294 (Dulverton) (1969) (provisional)
Sheet 295 (Taunton) (1984)
Inner Bristol Channel Sheets 263 and 279 (1995)
Geophysical information map

### Sections

Vertical sections No.47. 1873. Vertical Sections of the Lower Lias and Rhaetic or Penarth Beds of Glamorgan, Somerset, and Gloucester-shires. No.8. Watchet, Somerset, Railway Section.

### BGS Technical and other reports

GEOLOGY

EDWARDS, R A. 1996. Selected geological locality details for the Minehead district. *British Geological Survey Technical Report*, WA/96/32.

ENGINEERING GEOLOGY

FORSTER, A. 1995. The engineering geology of the Minehead area. 1:50 000 Geological Sheet 278. *British Geological Survey Technical Report*, WN/95/22.

BIOSTRATIGRAPHY

DAVEY, R J. 1976. Palynology of grab samples from the Bristol Channel. *Institute of Geological Sciences Report*, PDL 76/289.

HARLAND, R. 1974. Organic-walled microplankton from gravity core BC 126, Bristol Channel. *Institute of Geological Sciences Report*, PDL 74/27.

IVIMEY-COOK, H C. 1992. Late Triassic and Lower Jurassic fossils from Sheet 278 (Minehead). *British Geological Survey Technical Report*, WH/92/338R.

IVIMEY-COOK, H C. 1993. The Jurassic rocks in BGS offshore boreholes 72/63 and 73/62 [51/-04/6 and 51/-04/244]. *British Geological Survey Technical Report*, WH/93/5R.

IVIMEY-COOK, H C. 1993. The BGS boreholes Selworthy No.1 and Selworthy No.2 near Minehead. *British Geological Survey Technical Report*, WH/9314R.

IVIMEY-COOK, H C. 1993. The Penarth Group on the Minehead Sheet. *British Geological Survey Technical Report*, WH/93/269R.

IVIMEY-COOK, H C. 1993. The Lias Group on the Minehead Sheet. *British Geological Survey Technical Report*, WH/93/279R.

RIDING, J B, 1994. A palynological examination of four samples from the Bristol Channel. *British Geological Survey Technical Report*, WH/94/81R.

RILEY, N J. 1995. Foraminifera and algae from limestone clasts within the Permo-Trias of SW England. *British Geological Survey Technical Report*, WH/95/140R.

WARRINGTON, G. 1970. Report on material from the Minehead area (Sheet 278) submitted by D J C Laming for palynological examination. *Institute of Geological Sciences Report*, PDL 70/20.

WARRINGTON, G. 1974. Palynology report: CSU.1 samples, Bristol Channel area (S1.036.S). *Institute of Geological Sciences Report*, PDL 74/31.

WARRINGTON, G. 1993. Palynology report: the Mesozoic sequence of Selworthy No. 2 Borehole, a preliminary account with documentation of the Penarth Group and contiguous beds

(Triassic) (Sheet 278: Minehead). *British Geological Survey Technical Report*, WH/93/226R.

WARRINGTON, G. 1994.   Palynology report: the latest Triassic to earliest Jurassic sequence, Selworthy No. 2 Borehole (Sheet 278: Minehead).   *British Geological Survey Technical Report*, WH/94/166R.

WARRINGTON, G. 1994.   Palynology report: Mercia Mudstone Group (Triassic), between Porlock and Dunster, west Somerset (Sheet 278: Minehead).   *British Geological Survey Technical Report*, WH/94/293R.

WARRINGTON, G, and OWENS, B (compilers). 1977. Micropalaeontological biostratigraphy of offshore samples from south-west Britain.   *Report of the Institute of Geological Sciences*, No. 77/7.

WILKINSON, I P. 1994.   Jurassic microfaunas from a suite of samples off Minehead.   *British Geological Survey Technical Report*, WH/94/65R.

WILKINSON, I P, and HALLIWELL, G P (compilers). 1980. Offshore micropalaeontological biostratigraphy of southern and western Britain.   *Report of the Institute of Geological Sciences*, No. 79/9.

SEDIMENTOLOGY

JONES, N S. 1995.   The sedimentology of the Devonian Hangman Sandstone Formation and Permo-Triassic strata from the Minehead Sheet 278.   *British Geological Survey Technical Report*, WH/95/119R.

PETROGRAPHY

STRONG, G E. 1995.   Petrography of Permo-Triassic and Devonian rock specimens from the Minehead area, Somerset. *British Geological Survey Technical Report*, WG/95/10.

ROCK PROPERTIES

ENTWISLE, D C. 1995.   Density and porosity determinations on thirteen Devonian and Triassic rock samples from the Minehead area, Somerset.   *British Geological Survey Engineering Geology and Geophysics Group, Laboratory Report*, No. 95/3.

REMOTE SENSING

MARSH, S H. 1993.   Landsat TM Interpretation Notes: Minehead, SW England.   *British Geological Survey Remote Sensing Group. Project Note*, RSG/93/4.

GEOCHEMISTRY AND MINERALISATION

JONES, R C, BEER, K E, and TOMBS, J M C. 1987.   Geochemical and geophysical investigations in Exmoor and the Brendon Hills.   *British Geological Survey, Mineral Reconnaissance Programme Report*, No. 90.

## LIST OF BGS 1:10 000 GEOLOGICAL MAPS

The following is a list of the 1:10 000 scale geological maps included wholly, or partly, within the area of the 1:50 000 scale Minehead (278) geological sheet, with the initials of the surveyors and the date of survey of each map. The surveyors were R A Edwards, A Whittaker, J C Thackray and E A Edmonds. Copies of the manuscript National Grid maps are deposited for public reference in the libraries of the British Geological Survey at Keyworth and Exeter. Uncoloured dyeline copies of these maps

are available for purchase from the British Geological Survey, Keyworth, Nottingham NG12 5GG. Partially surveyed sheets are marked by an asterisk.

| SS 74 NE | Brendon | AW, RAE | 1973, 1994 |
| SS 74 SE | Exe Head | AW, EAE, RAE | 1973–76, 1993 |
| SS 84 NW | Culbone and Oare | RAE | 1993 |
| SS 84 NE | Porlock | RAE | 1993–94 |
| SS 84 SW* | Larkbarrow | RAE | 1993 |
| SS 84 SE* | Stoke Pero | RAE | 1993 |
| SS 94 NW | Selworthy | RAE | 1994 |
| SS 94 NE | Minehead | RAE | 1994 |
| SS 94 SW* | Wootton Courtenay | RAE | 1994 |
| SS 94 SE* | Dunster | RAE | 1994 |
| ST 04 SW* | Blue Anchor | RAE, AW | 1970, 1994 |
| ST 04 SE* | Watchet and Williton | AW, JCT, RAE | 1970, 1975, 1994 |

### History of survey

The district covered by the 1:50 000 Series Minehead Sheet (278) forms part of the region covered by Old Series one-inch Sheet 20, with a small area west of Oare falling within Old Series Sheet 27. These sheets were surveyed geologically on the one-inch scale by Sir Henry T De la Beche; Sheet 20 was issued in 1834 and revised in 1839. The classic 'Report on the geology of Cornwall, Devon and West Somerset' (De la Beche, 1839) was the first memoir of the Geological Survey.

A one-inch resurvey of the district was begun in the Watchet area by Mr J H Blake in 1872–73. Mr W A E Ussher completed the survey of the remainder of the district in 1874 and 1879, also on the one-inch scale, and carried out some revision of Blake's work. However, no New Series one-inch map of the district was published, and consequently, the new 1:50 000 scale map is the first official geological map of the district to appear for over 150 years. In the absence of the New Series one-inch map, no memoir was issued, although Mr Ussher prepared manuscript notes for a memoir.

The primary 1:10 000 scale survey of the district was carried out by Dr R A Edwards in 1993–94. The survey was extended into the area of the Dulverton (294) sheet as far south as the National Grid Northing [1]43 and east of the National Grid Easting [3]00 (near Dunster), in order to include the coast between Blue Anchor and Watchet. Dr A Whittaker mapped an overlap about 1 km deep into the western side of the district, as part of the survey of the Ilfracombe district (Sheet 277) in 1973. He also mapped the foreshore exposures of the Late Triassic and Early Jurassic strata between Blue Anchor and Watchet, using overlays on 1:5000 scale vertical aerial photographs, as part of the survey of the Weston-super-Mare district (Sheet 279) in 1970 and 1975.

The geological map of the offshore area was prepared in 1993–95 by Dr C D R Evans and Mr D H Jeffery of the Coastal Geology Group of BGS.

### LIST OF MAIN BOREHOLES

ONSHORE

Selworthy No.1 [SS 9244 4630] (SS 94 NW/2).
Selworthy No.2 [SS 9244 4618] (SS 94 NW/3).
Holloway Street, Minehead [SS 9673 4637] (SS 94 NE/37).
Butlin's 'Somerwest World' Holiday Camp, Minehead (SS 94 NE/8, 9, 29 to 34).

## OFFSHORE

Two boreholes 51/-04/6 [SS 7880 5456] and 51/-04/244 [SS 9173 5513] (also known using the annual numbering system as 72/63 and 73/62) were drilled by BGS using the drill vessel mv *Whitethorn* in the centre of the Bristol Channel Syncline.

## SOURCES AND TYPES OF INFORMATION

### GEOPHYSICS

Gravity and aeromagnetic data are held digitally in the National Gravity Databank and the National Aeromagnetic Databank at BGS Keyworth.

The results of the gravity survey of the Porlock Basin are held digitally at BGS Keyworth.

Shallow seismic profiles were run in the offshore part of the district by Sonarmarine Ltd in 1979 (under contract to BGS). The seismic grid, run using a boomer high-resolution system, extended westwards to Hurlstone Point. Additional lines across the western part of the area, run by BGS in 1981 using a sparker and airgun system, were of poor quality and of little value in interpreting the structure of the area. The data from both surveys are held at BGS Keyworth.

Five deep seismic reflection profiles in the offshore part of the district come from a variety of non-exclusive proprietary surveys owned by Geco-Prakla, a division of Geco Geophysical Company Limited, and that company should be approached for permission to use the data. The profiles were run in 1985 by Merlin Profilers Ltd..

### OFFHORE SAMPLES

Offshore samples collected by BGS, mostly obtained using a gravity corer during cruises in 1976 and 1979, are held at BGS Keyworth. Micropalaeontological and other data from these samples were published in Warrington and Owens (1977) and Wilkinson and Halliwell (1980). BGS has none of the original data described by Lloyd et al. (1973) and Evans and Thompson (1979). Data from gravity core samples were obtained in the 1960s by University College London and Institute of Oceanographic Sciences workers (Lloyd et al., 1973). More detailed work was carried out by workers from Swansea University in the early 1970s (Evans, 1973; Evans and Thompson, 1979) using gravity corer, seismic profiling and side scan sonar. Details of the taxa recovered from the samples are given in Evans (1973). The original samples and residues do not appear to have been retained.

### HYDROGRAPHIC DATA

Hydrographic and sidescan sonar data for the offshore area were obtained from detailed surveys carried in the 1980s by the Hydrographic Office of the Ministry of Defence and are held in the Hydrographic Office database.

### GEOCHEMICAL DATA

The results of analyses of four samples of Luccombe Breccia for Au are held in the BGS Geochemical Database at BGS Keyworth.

### REMOTE SENSING DATA

Landsat data from TM scene 203–024, acquired on 30 October 1988, are available for the district. About 25 per cent of the district has variable cloud cover on this scene. The data are held at BGS Keyworth.

### HYDROGEOLOGY

Data on water boreholes, wells and springs are held in the BGS database at Wallingford. The Environment Agency also holds data on groundwater sources. At BGS Wallingford, information is held on 46 groundwater sources, comprising 26 boreholes or wells and 20 springs that have been used for water supplies.

### WASTE DISPOSAL

Details of licensed landfill sites are maintained by Somerset County Council. The Environment Agency has information on some sites.

### RADON

The results of the assessment of the relationship between geology and radon potential in Somerset were published by Appleton and Ball (1995). Further information may be obtained from BGS Keyworth. The National Radiological Protection Board (NRPB) will carry out house surveys for radon in designated 'Affected Areas'.

### SSSIs AND RIGS

Maps showing the boundaries of Sites of Special Scientific Interest (SSSIs) and citation sheets for each site are held by English Nature. SSSI boundary data are also available in digital form. There are two geological SSSIs in the district. The Glenthorne SSSI [SS 794 499 to 805 495], near the western boundary of the district, covers outcrops of Hangman Sandstone. In the east of the district, coastal exposures to the east of Blue Anchor [ST 033 435] form part of a geological SSSI in Late Triassic and Early Jurassic rocks which extends for about 19 km eastwards to Lilstock [ST 195 462].

A list of Regionally Important Geological/Geomorphological Sites (RIGS) in the district, with boundaries and citation information, is available by application to the Manager, Somerset Environmental Records Centre, Taunton.

### PHOTOGRAPHS

Copies of photographs taken in the district area are deposited for reference at BGS Keyworth and belong to Series A and GS. Prints, and in some cases slides, may be purchased (prices, and details of formats and subjects, are available on application to BGS Keyworth).

### THIN SECTIONS

A total of 29 thin sections of rocks from the Hangman Sandstone and various Permo-Triassic formations are held in the England and Wales Sliced Rocks collection at BGS Keyworth. Charges and conditions of access to the collection are available on request from BGS Keyworth.

### FOSSILS

Macrofossils and micropalaeontological residues for samples collected from the district are held at BGS Keyworth. For the onshore area, three palynological residues collected in 1970 are held in the SAL numbering system, and 66 later residues are held in the MPA numbering system; 53 of these are from the

Selworthy No. 2 Borehole [SS 9244 4618] and the remainder from outcrop. Residues for offshore samples are stored at BGS Keyworth in the CSA numbering system.

The Devonian Hangman Sandstone is almost barren of fossils, and there is only one plant specimen in the BGS collections. The foreshore and cliff sections in Triassic and Jurassic rocks between Blue Anchor and Watchet have not been systematically collected during the recent survey, and there are only a few Late Triassic–Early Jurassic specimens from there in the collections. Macrofossils from the Triassic and Jurassic rocks in the Selworthy No. 2 Borehole are registered in the BDE and BKE series. Enquiries concerning all macrofossil material should be directed to the Curator, Biostratigraphy Collections, BGS Keyworth.

BOREHOLE DATA

Access to a computerised Borehole Index, showing the availability of boreholes in the BGS borehole database, is available free of charge at BGS Keyworth. Paper copies of borehole and trial pit logs from the district are held at the BGS South-Western England Office at Exeter. Prices for copies of borehole or trial pit logs, fees for inspection of brehole logs, and prices for value-added information (e.g. additional information or classifications derived from BGS-examined borehole material) are available from the BGS Exeter Office.

Core samples from the Selworthy boreholes and offshore boreholes 72/63 and 73/62 are curated at BGS Keyworth.

The BGS holds a large number of site investigation reports in the Keyworth and Exeter offices. These may be consulted subject to the owners' permission and BGS access arrangements.

AERIAL PHOTOGRAPHS

Monochrome aerial photographs supplied by the Ordnance Survey at a scale of about 1:24 800 and flown in 1971 were used during the survey of the main part of the district. A set of monochrome aerial photographs at about 1:5000 scale between Porlock and Watchet form part of a special survey of the coastal zone specially flown for the (then) IGS in 1967 by Fairey Surveys Ltd.. Both sets of photographs are held at the BGS Exeter Office.

BGS LEXICON OF NAMED ROCK UNIT DEFINITIONS

Definitions of the named rock units shown on BGS maps, including those shown on the 1:50 000 Series Minehead Sheet 278 are held in the Lexicon database. Information on how to consult the database can be obtained from the Lexicon Manager at BGS Keyworth.

BGS PETMIN DATABASE

A database of thin sections and rock samples is maintained by the Mineralogy and Petrology Group at BGS Keyworth. The Group Manager at BGS Keyworth should be contacted for further information, including methods of accessing the database.

PLANNING DATA

The following planning documents are relevant to the district. Copies may be obtained from Somerset County Council, West Somerset District Council, or Exmoor National Park Authority.

Somerset Structure Plan, Alteration No. 2, 1993 (for period 1993–2001).

Somerset Structure Plan Review, Consultation Draft, 1995 (for period to 2011).
Minehead Area Local Plan, 1989 (for period 1989–1996).
West Somerset District Local Plan (Consultative Report), 1995 (for period to 2006).
Exmoor National Park Local Plan, 1994 (for period 1994–2001) (also contains Minerals Local Plan and Waste Local Plan for the Park).

ADDRESSES FOR DATA SOURCES AND SUMMARY OF MAIN SERVICES AND PRODUCTS AVAILABLE

British Geological Survey, Headquarters, Keyworth, Nottingham NG12 5GG.
*Telephone* 0115 936 3100.    *Fax* 0115 936 3200.
**Enquiry service; 1:10 000 maps (sale and reference copies); borehole samples and offshore samples; geophysical data and seismic data; geochemical data; remote sensing data; radon; fossils; thin sections; Petmin database; Lexicon; library; publications sales desk.**

British Geological Survey, South-west England Regional Office, St Just, 30 Pennsylvania Road, Exeter, Devon EX4 6BX.
*Telephone* 01392 278312.    *Fax* 01392 437505.
**Enquiry service; 1:10 000 maps (reference copies); borehole and trial pit records; library; publication sales (south-west England area); aerial photographs.**

British Geological Survey, Hydrogeology Group, Maclean Building, Crowmarsh Gifford, Wallingford, Oxfordshire OX10 8BB.
*Telephone* 01491 838800.    *Fax* 01491 692345.
**Enquiry service; wells, springs and water borehole records.**

Environment Agency, South Western Region, Rivers House, East Quay, Bridgwater, Somerset TA6 4YS.
*Telephone* 01278 457333.    *Fax* 01278 452985.
**Groundwater; flood risk; landfill sites.**

Somerset County Council, County Hall, Taunton, Somerset TA1 4DY.
*Telephone* 01823 333451.    *Fax* 01823 332773
(Environment).
**Planning; landfill sites.**

West Somerset District Council, Council Offices, Williton, Somerset TA4 4QA.
*Telephone* 01984 632291.    *Fax* 01984 633022.
**Planning.**

Somerset Environmental Records Centre (SERC), Pickney, Kingstone St Mary, Taunton, Somerset TA2 8AS.
*Telephone* 01823 451778.
**RIGS.**

Exmoor National Park Authority, Exmoor House, Dulverton, Somerset TA 22 9HL.
*Telephone* 01398 23665.    *Fax* 01398 23150.
**Planning, conservation.**

English Nature, Somerset and Avon Team, Roughmoor, Bishop's Hull, Taunton, Somerset TA1 5AA.
*Telephone* 01823 283211.    *Fax* 01823 272978.
**Conservation; SSSIs.**

National Radiological Protection Board, Chilton, Didcot, Oxfordshire OX11 0RQ.
*Telephone* 01235 831600.    *Fax* 01235 833891.
**Radon.**

# REFERENCES

Most of the references listed below are held in the Library of the British Geological Survey at Keyworth, Nottingham. Copies of the references can be purchased subject to current copyright legislation.

AL-SAADI, H M. 1967. A gravity investigation of the Pickwell Down Sandstone, north Devon. *Geological Magazine*, Vol. 104, 63–72.

ANDERTON, R, BRIDGES, P H, LEEDER, M R, and SELLWOOD, B W. 1979. *A dynamic stratigraphy of the British Isles.* (London: George Allen and Unwin.)

ANON. 1981. British Standard code of practice for site investigation, B.S. 5930. British Standards Institution. (London: HMSO.)

ANON. 1992. Study of landslipped coastal slopes and woodland — Culbone Woods, Somerset. First report to Exmoor National Park by Integral Geotechnique Ltd.

ANON. 1994. Study of landslipped coastal slopes and woodland — Culbone Woods, Somerset. Second report to Exmoor National Park by Integral Geotechnique Ltd.

APPLETON, J D, and BALL, T K. 1995. Radon and background radioactivity from natural sources: characteristics, extent and relevance to planning and development in Great Britain. *British Geological Survey Technical Report*, WP/95/2.

BARRIE, J V. 1980. Heavy mineral distribution in bottom sediments of the Bristol Channel, U.K. *Estuarine and Coastal Marine Science*, Vol. 11, 369–381.

BASSETT, M G, and COPE, J C W. 1993. Discussion on new evidence for a major geological boundary at shallow depth, N. Devon. *Journal of the Geological Society of London*, Vol. 150, 1197–1199.

BASSETT, M G, BLUCK, B J, CAVE, R, HOLLAND, C H, and LAWSON, J D. 1992. Silurian. 37–56 *in* Atlas of palaeogeography and lithofacies. COPE, J C W, INGHAM, J K, and RAWSON, P F (editors). *Memoir of the Geological Society of London*, No. 13.

BATTIAU-QUENEY, Y. 1984. The pre-glacial evolution of Wales. *Earth Surface and Landforms*, Vol. 9, 229–252.

BAYERLEY, M, and BROOKS, M. 1980. A seismic study of deep structure in South Wales using quarry blasts. *Geophysical Journal of the Royal Astronomical Society*, Vol. 60, 1–19.

BEACH, A. 1987. A regional model for linked tectonics in north-west Europe. 43–48 in *Petroleum geology of North West Europe, Volume 1*. BROOKS, J, and GLENNIE, K W (editors). (London: Graham and Trotman.)

BEER, K E. 1988. Addenda and corrigenda to the 'Metalliferous mining region of south-west England' (Dines, H G, 1956). *Economic Memoirs of the Geological Survey of Great Britain.*

BENTON, M J, WARRINGTON, G, NEWELL, A J, and SPENCER, P S. 1994. A review of the British Middle Triassic tetrapod assemblages. 131–160 in *In the shadow of the dinosaurs: early Mesozoic tetrapods*. FRASER, N C, and SUES, H-D (editors). (New York: Cambridge University Press.)

BEVINS, R E, BLUCK, B J, BRENCHLEY, P J, FORTEY, R A, HUGHES, C P, INGHAM, J K, and RUSHTON, A W A. 1992. Ordovician. 19–36 *in* Atlas of palaeogeography and lithofacies. COPE, J C W, INGHAM, J K, and RAWSON, P F (editors). *Memoir of the Geological Society of London*, No. 13.

BLUCK, B J, COPE, J C W, and SCRUTTON, C T. 1992. Devonian. 57–66 *in* Atlas of palaeogeography and lithofacies. COPE, J C W, INGHAM, J K, and RAWSON, P F (editors). *Memoir of the Geological Society of London*, No. 13.

BOOMER, I D. 1991. Lower Jurassic ostracod biozonation of the Mochras Borehole. *Journal of Micropalaeontology*, Vol. 9, 205–218.

BOTT, M H P, DAY, A A, and MASSON SMITH, D. 1958. The geological interpretation of gravity and magnetic surveys in Devon and Cornwall. *Philosophical Transactions of the Royal Society*, Vol. 251. Series A. 161–191.

BOWN, P R. 1987. Taxonomy, evolution, and biostratigraphy of Late Triassic–Early Jurassic calcareous nanofossils. *Special Papers in Palaeontology*, Vol. 38, 1–117.

BOYD-DAWKINS, W B. 1864. Outline of the Rhaetic Formation of west and central Somerset. *Geological Magazine*, Vol. 1, 257–260.

BRADSHAW, R, and HAMILTON, D. 1967. Conjugate gypsum veins at Blue Anchor Point, Somerset. *Proceedings of the Bristol Natural History Society*, Vol. 31, 305–309.

BRASIER, M D, INGHAM, J K, and RUSHTON, W A. 1992. Cambrian. 13–18 *in* Atlas of palaeogeography and lithofacies. COPE, J C W, INGHAM, J K, and RAWSON, P F (editors). *Memoir of the Geological Society of London*, No. 13.

BRIDGE, J S. 1985. Palaeochannel patterns inferred from alluvial deposits: A critical evaluation. *Journal of Sedimentary Petrology*, Vol. 55, 579–589.

BRIDGE, J S, and LEEDER, M R. 1979. A simulation model of alluvial stratigraphy. *Sedimentology*, Vol. 26, 617–644.

BRISTOW, H W, and ETHERIDGE, R. 1873. Vertical sections No. 47. Lower Lias and Rhaetic Beds of Glamorgan-, Somerset-, and Gloucester-shire. No. 8. Watchet.

BRITISH GEOLOGICAL SURVEY. 1982. Hydrogeological map of the Permo–Trias and other minor aquifers of South West England. 1:100 000. (Keyworth: British Geological Survey.)

BRITISH GEOLOGICAL SURVEY. 1985. Pre-Permian geology of the United Kingdom (South). 1: 1 000 000. Two maps commemorating the 150th Anniversary of the British Geological Survey. (Keyworth, Nottingham: British Geological Survey on behalf of the Petroleum Engineering Division of the Department of Energy.)

BROOKS, M. 1987. Geophysical investigations in the Bristol Channel area (extended abstract). *Proceedings of the Geologists' Association*, Vol. 97, 397–398.

BROOKS, M, and THOMPSON, M S. 1973. The geological interpretation of a gravity survey of the Bristol Channel. *Journal of the Geological Society of London*, Vol. 129, 245–274.

BROOKS, M, and JAMES, D G. 1975. The geological results of seismic refraction surveys in the Bristol Channel, 1970–1973. *Journal of the Geological Society of London*, Vol. 131, 163–182.

BROOKS, M, and AL-SAADI, R H. 1977. Seismic refraction studies of geological structure in the inner part of the Bristol Channel. *Journal of the Geological Society of London*, Vol. 133, 433–445.

BROOKS, M, BAYERLEY, M, and LLEWELLYN, D J. 1977. A new geological model to explain the gravity gradient across Exmoor, north Devon. *Journal of the Geological Society of London*, Vol. 133, 385–393.

BROOKS, M, TRAYNER, P M, and TRIMBLE, T J. 1988. Mesozoic reactivation of Variscan thrusting in the Bristol Channel area, UK. *Journal of the Geological Society of London*, Vol. 145, 439–444.

BROOKS, M, HILLIER, B V, and MILIORIZOS, M. 1993. New seismic evidence for a major geological boundary at shallow depth, N. Devon. *Journal of the Geological Society of London*, Vol. 150, 131–135.

BROOKS, M, MILIORIZOS, M, and HILLIER, B V. 1994. Deep structure of the Vale of Glamorgan, South Wales, UK. *Journal of the Geological Society of London*, Vol. 151, 909–917.

CHADWICK, R A, KENOLTY, N, and WHITTAKER, A. 1983. Crustal structure beneath southern England from deep seismic profiles. *Journal of the Geological Society of London*, Vol. 140, 893–911.

CHAMPERNOWNE, A, and USSHER, W A E. 1879. Notes on the structure of the Palaeozoic districts of west Somerset. *Quarterly Journal of the Geological Society of London*, Vol. 25, 532–548.

CIFELLI, R. 1959. Bathonian foraminifera of England. *Bulletin of the Museum of Comparative Zoology at Harvard College*, Vol. 121, 265–368.

CLARE, R. 1987. An introduction to the work of the Severn Tidal Power Group 1983–1985. 3–16 in *Tidal Power: proceedings of the symposium organised by the Institution of Civil Engineers, London, 30–31 October 1986*. (London: Thomas Telford.)

CLARKE, F J P. 1982. Severn Barrage schemes from 1849 onwards. 3–8 in *Severn Barrage: proceedings of a symposium organised by the Institution of Civil Engineers, London, 8–9 October 1981*. (London: Thomas Telford.)

CLARKE, R H, and SOUTHWOOD, T R E. 1989. Risks from ionising radiation. *Nature, London*, Vol. 338, 197–198.

COLLINS, M B. 1989. Sediment fluxes in the Bristol Channel. *Proceedings of the Ussher Society*, Vol. 7, 107–111.

COLLINSON, J D, and THOMPSON, D B. 1989. *Sedimentary structures* (2nd edition). (London: Unwin Hyman.)

COOK, A H, and THIRLAWAY, H I S. 1952. A gravimeter survey in the Bristol and Somerset coalfields. *Quarterly Journal of the Geological Society of London*, Vol. 107, 255–285.

COOPER, J A G, ORFORD, J D, MCKENNA, J, JENNINGS, S C, SCOTT, B M, and MALVAREZ, G. 1995. Meso-scale behaviour of Atlantic coastal systems under secular climatic change and sea-level rise. Contribution to *The Impacts of climate change and relative sea-level rise on the environmental resources of European coasts*, CEC Environment and Climate Programme (Third Framework), No. EV5V-CT93-0258.

COPE, J C W. 1987. The Pre-Devonian geology of south-west England. *Proceedings of the Ussher Society*, Vol. 7, 468–473.

COPE, J C W, and BASSETT, M G. 1987. Sediment sources and Palaeozoic history of the Bristol Channel area. *Proceedings of the Geologists' Association*, Vol. 98, 315–320.

COPE, J C W, GETTY, T A, HOWARTH, M K, MORTON, N, and TORRENS, H S. 1980. A correlation of Jurassic rocks in the British Isles. Part One: Introduction and Lower Jurassic. *Memoir of the Geological Society of London*, No. 14.

COPESTAKE, P. 1989. Triassic. 97–124 in *Stratigraphical atlas of fossil foraminifera* (2nd edition). JENKINS, D G, and MURRAY, J W (editors). (Chichester: Ellis Horwood.)

CORDEY, W G. 1963. Oxford Clay foraminifera from England (Dorset–Northamptonshire) and Scotland. Unpublished PhD thesis, University of London.

CORNFORD, C. 1986. The Bristol Channel Graben: organic geochemical limits on subsidence and speculation on the origin of inversion. *Proceedings of the Ussher Society*, Vol. 6, 360–367.

CORNWELL, J D. 1986. The geological significance of some geophysical anomalies in western Somerset. *Proceedings of the Ussher Society*, Vol. 6, 383–388.

COX, F C, DAVIES, J R, and SCRIVENER, R C. 1986. *The distribution of high grade sandstone for aggregate usage in parts of south west England and South Wales*. Report for the Department of the Environment. (Keyworth, Nottingham: British Geological Survey.)

CULLINGFORD, R A. 1982. The Quaternary. 249–290 in *The geology of Devon*. DURRANCE, E M, and LAMING, D J C (editors). (Exeter: University of Exeter.)

CULVER, S J, and BANNER, F T. 1978. The significance of derived pre-Quaternary foraminifera in Holocene sediments of the north-central Bristol Channel. *Marine Geology*, Vol. 29, 187–207.

DART, C J, MCCLAY, K, and HOLLINGS, P N. 1995. 3D analysis of inverted extensional fault systems, southern Bristol Channel, UK. 393–413 in Basin inversion. BUCHANAN, J G, and BUCHANAN, P G (editors). *Special Publication of the Geological Society of London*, No. 88.

DAVISON, I. 1994. Structural field trip guide to the Watchet–Lilstock area, N. Somerset, No. 41. Petroleum Exploration Society of Great Britain.

DE LA BECHE, H T. 1839. *Report on the geology of Cornwall, Devon, and West Somerset*. (London: Longman, Orme, Brown, Green and Longmans.)

DINES, H G. 1956. The metalliferous mining region of south-west England. *Memoir of the Geological Survey of Great Britain*.

DODSON, M H, and REX, D C. 1971. Potassium-argon ages of slates and phyllites from South West England. *Quarterly Journal of the Geological Society of London*, Vol. 126, 465–499.

DONOVAN, D T, and STRIDE, A A. 1961. Erosion of a rock floor by tidal sand streams. *Geological Magazine*, Vol. 98, 393–398.

EAGAR, R M C, BAINES, J G, COLLINSON, J D, HARDY, P G, OKOLO, S A, and POLLARD, J E. 1985. Trace fossil assemblages and their occurrence in Silesian (mid-Carboniferous) deltaic sediments of the central Pennine basin, England. 99–149 in Biogenic structures: their use in interpreting depositional environments. CURRAN, H A (editor). *Special Publication of the Society of Economic Paleontologists and Mineralogists*, No. 35.

EDMONDS, E A, and WILLIAMS, B J. 1985. Geology of the country around Taunton and the Quantock Hills. *Memoir of the British Geological Survey*, Sheet 295 (England and Wales).

EDMONDS, E A, WHITTAKER, A, and WILLIAMS, B J. 1985. Geology of the country around Ilfracombe and Barnstaple. *Memoir of the British Geological Survey*, Sheets 277 and 293 (England and Wales).

EDWARDS, R A. 1996. Selected geological locality details for the Minehead district. *British Geological Survey Technical Report*, WA/96/32.

EDWARDS, R A, WARRINGTON, G, SCRIVENER, R C, JONES, N S, HASLAM, H W, and AULT, L. 1997. The Exeter Group, south Devon, England: a contribution to the early post-Variscan stratigraphy of NW Europe. *Geological Magazine,* Vol. 134, 177–197.

EDWARDS, R A, and SCRIVENER, R C. *In press.* Geology of the country around Exeter. *Memoir of the British Geological Survey,* Sheet 325 (England and Wales).

ETHERIDGE, R. 1867. On the physical structure of west Somerset and north Devon and on the palaeontological value of the Devonian fossils. *Quarterly Journal of the Geological Society of London,* Vol. 23, 568–698.

ETHERIDGE, R. 1872. Notes upon the physical structure of the Watchet area, and the relation of the Secondary rocks to the Devonian Series of west Somerset. *Proceedings of the Cotteswold Naturalists' Field Club,* Vol.6, 35–48.

EVANS, C D R. 1982. The geology and superficial sediments of the inner Bristol Channel and Severn Estuary. 35–42 in *Severn Barrage: proceedings of a symposium organised by the Institution of Civil Engineers, London, 8–9 October 1981.* (London: Thomas Telford.)

EVANS, D J. 1973. The stratigraphy of the central part of the Bristol Channel. Unpublished PhD thesis, University of Wales (Swansea).

EVANS, D J, and THOMPSON, M S. 1979. The geology of the central Bristol Channel and the Lundy area, South Western Approaches, British Isles. *Proceedings of the Geologists' Association,* Vol. 90, 1–14.

EVANS, J W. 1922. The geological structure of the country around Combe Martin, north Devon. *Proceedings of the Geologists' Association,* Vol. 33, 201–234.

EVANS, K M. 1983. Note on the age and fauna of the Lynton Beds (Lower Devonian) of north Devon. *Geological Journal,* Vol. 18, 297–305.

FINDLAY, D C, COLBORNE, G J N, COPE, D W, HARROD, T R, HOGAN, D V, and STAINES, S J. 1984. Soils and their use in south-west England. *Bulletin of the Soil Survey of England and Wales,* No. 14.

FIRMAN, R J. 1984. A geological approach to the history of English alabaster. *Mercian Geologist,* Vol. 9, 161–178.

FIRMAN, R J. 1989. Alabaster update — research in progress. *Mercian Geologist,* Vol. 12, 63–70.

FOLK, R L, and WARD, W C. 1957. Brazos River bar, a study in the significance of grain-size parameters. *Journal of Sedimentary Petrology,* Vol. 27, 3–27.

GILBERTSON, D D, and MOTTERSHEAD, D N. 1975. The Quaternary deposits at Doniford, west Somerset. *Field Studies,* Vol. 4, 117–129.

GODWIN-AUSTEN, R A C. 1865. On the submerged forest-beds of Porlock Bay. *Quarterly Journal of the Geological Society of London,* Vol. 22, 1–9.

GRAINGER, P, and KALAUGHER, P G. 1996. Photographic monitoring of cliff recession at Watchet, Somerset. *Proceedings of the Ussher Society,* Vol. 9, 17–20.

GREEN, G W, and WELCH, F B A. 1965. Geology of the country around Wells and Cheddar. *Memoir of the Geological Survey of Great Britain,* Sheet 280 (England and Wales).

GRINSELL, L V. 1970. *The archaeology of Exmoor.* (Newton Abbot: David and Charles.)

HALLAM, A. 1964. Origin of the limestone–shale rhythm in the Blue Lias of England: a composite theory. *Journal of Geology,* Vol. 72, 157–169.

HAMILTON, G B. 1982. Triassic and Jurassic calcareous nannofossils. 17–39 in *A stratigraphical index of calcareous nannofossils.* LORD, A R (editor). (Chichester: Ellis Horwood.)

HAWKINS, A B. 1971. The late Weichselian and Flandrian Transgression of south west Britain. *Quaternaria,* Vol. 14, 115–130

HEYWORTH, A, and KIDSON, C. 1982. Sea-level changes in southwest England and Wales. *Proceedings of the Geologists' Association,* Vol. 93, 91–111

HODGES, P. 1994. The base of the Jurassic System: new data on the first appearance of *Psiloceras planorbis* in southwest Britain. *Geological Magazine,* Vol. 131, 841–844.

HOGG, S E. 1982. Sheetfloods, sheetwash, sheetflow, or…? *Earth-Science Reviews,* Vol. 18, 59–76.

HORNER, L. 1816. Sketch of the geology of the south-western part of Somersetshire. *Transactions of the Geological Society of London,* Vol. 3, 338–384.

INSTITUTE OF GEOLOGICAL SCIENCES. 1974. IGS boreholes 1973. *Report of the Institute of Geological Sciences,* No. 74/7.

IVIMEY-COOK, H C. 1993. The Jurassic rocks in BGS offshore boreholes 72/63 and 73/62 [51/-04/6 and 51/-04/244]. *British Geological Survey Technical Report,* WH/93/5R.

IVIMEY-COOK, H C, and DONOVAN, D T. 1983. Appendix 3. The fauna of the Lower Jurassic. 126–130 in Geology of the country around Weston-super-Mare. WHITTAKER, A, and GREEN, G W. *Memoir of the Geological Survey of Great Britain,* Sheet 279, with parts of sheets 263 and 295 (England and Wales).

JONES, N S. 1995. The sedimentology of the Devonian Hangman Sandstone Formation and Permo-Triassic strata from the Minehead Sheet 278. *British Geological Survey Technical Report,* WH/95/119R.

JONES, R C, BEER, K E, and TOMBS, J M C. 1987. Geochemical and geophysical investigations in Exmoor and the Brendon Hills. *British Geological Survey, Mineral Reconnaissance Programme Report,* No. 90

KAMERLING, P. 1979. The geology and hydrocarbon habitat of the Bristol Channel Basin. *Journal of Petroleum Geology,* Vol. 2, 75–93.

KELLAWAY, G A, and WELCH, F B A. 1993. Geology of the Bristol district. *Memoir of the British Geological Survey,* Bristol district special sheet (England and Wales).

KIDSON C. 1977. The coast of South West England. 257–298 in *The Quaternary history of the Irish Sea.* KIDSON, C, and TOOLEY, M J (editors). (Liverpool: Seel House Press.)

KIDSON, C, and HEYWORTH, A. 1976. The Quaternary deposits of the Somerset levels. *Quarterly Journal of Engineering Geology,* Vol. 9, 217–235.

KIRBY, R. 1988. Sedimentological implications of building the Cardiff–Weston barrage in the Severn Estuary. *Proceedings of the Ussher Society,* Vol. 7, 13–17.

KNIGHT, R R W. 1990. The Devonian of north Devon — a palynologist's (or 'conodontologist's') dream or nightmare? [Abstract]. *Proceedings of the Ussher Society,* Vol. 7, 306.

KRUMBEIN, W C. 1934. Size frequency distribution of sediments. *Journal of Sedimentary Petrology,* Vol.4, 65–77.

LAMING, D J C. 1968. New Red Sandstone stratigraphy in Devon and West Somerset. *Proceedings of the Ussher Society,* Vol. 2, 23–25.

LANE, R. 1965. The Hangman Grits — an introduction and stratigraphy. *Proceedings of the Ussher Society,* Vol. 1, 166–167.

LESLIE, A B, SPIRO, B, and TUCKER, M E. 1993. Geochemical and mineralogical variations in the upper Mercia Mudstone Group (Late Triassic), southwest Britain: correlation of outcrop sequences with borehole geophysical logs. *Journal of the Geological Society of London*, Vol. 150, 67–75.

LEVINSON, A A. 1974. *Introduction to exploration geochemistry.* (Calgary: Applied Publishing.)

LLOYD, A J, SAVAGE, R J D, STRIDE, A A, and DONOVAN, D T. 1973. The geology of the Bristol Channel floor. *Philosophical Transactions of the Royal Society, A*, Vol. 274, 595–626.

LORD, A R, and BOOMER, I D. 1990. The occurrence of ostracods in the Triassic/Jurassic boundary interval. *Les Cahiers de l'Université Catholique de Lyon, Série Sciences*, Vol. 3, 119–126.

LOTT, G K, SOBEY, R A, WARRINGTON, G, and WHITTAKER, A. 1982. The Mercia Mudstone Group (Triassic) in the western Wessex Basin. *Proceedings of the Ussher Society*, Vol. 5, 340–346.

MACQUAKER, J H S. 1984. Diagenetic modifications of primary sedimentological fabric in the Westbury Formation (Upper Triassic) of St Audrie's Bay, north Somerset. *Proceedings of the Ussher Society*, Vol. 6, 95–99.

MACQUAKER, J H S, FARRIMOND, P, and BRASSELL, S C. 1986. Biological markers in the Rhaetian black shales of South West Britain. *Organic Geochemistry*, Vol. 10, 93–100.

MADDOX, S J, BLOW, R, and HARDMAN, M. 1995. Hydrocarbon prospectivity of the Central Irish Sea Basin with reference to Block 42/12, offshore Ireland. 59–77 *in* The petroleum geology of Ireland's offshore basins. CROCKE, P F, and SHANNON, P M (editors). *Special Publication of the Geological Society of London*, No. 93.

MARTIN, E C. 1909. The probable source of the limestone pebbles in the Bunter Conglomerate of west Somerset. *Geological Magazine*, Vol. 46, 160–165.

MAYALL, M J. 1979. The clay mineralogy of the Rhaetic transgression in Devon and Somerset — environmental and stratigraphical implications. *Proceedings of the Ussher Society*, Vol. 4, 303–311.

MAYALL, M J. 1981. The Late Triassic Blue Anchor Formation and the initial Rhaetian marine transgression in south-west Britain. *Geological Magazine*, Vol. 118, 377–384.

MAYALL, M J. 1983. An earthquake origin for synsedimentary deformation in a late Triassic (Rhaetian) lagoonal sequence, southwest Britain. *Geological Magazine*, Vol. 120, 613–622.

McGEE, W J. 1897. Sheetflood erosion. *Bulletin of the Geological Society of America*, Vol. 8, 87–112.

McLAREN, P, COLLINS M B, GAO, S, and POWYS, R I L. 1993. Sediment dynamics of the Severn Estuary and inner Bristol Channel. *Journal of the Geological Society of London*, Vol. 150, 589–603.

McLAREN, P, and COLLINS, M B. 1989. *Sediment transport pathways in the Severn Estuary and Bristol Channel.* Severn Barrage Development Project, Ref. STPG/GEOSEA/3.1(iii)f.

MECHIE, J, and BROOKS, M. 1984. A seismic study of deep geological structure in the Bristol Channel area. *Geophysical Journal of the Royal Astronomical Society*, Vol. 78, 661–689.

NEMČOK, M, GAYER, R, and MILIORIZOS, M. 1995. Structural analysis of the inverted Bristol Channel Basin: implications for the geometry and timing of fracture porosity. 355–392 *in* Basin inversion. BUCHANAN, J G, and BUCHANAN, P G (editors). *Special Publication of the Geological Society of London*, No. 88.

NICKLESS, E F P, BOOTH, S J, and MOSLEY, P N. 1976. The celestite resources of the area north-east of Bristol, with notes on occurrences north and south of the Mendip Hills and the Vale of Glamorgan. *Mineral Assessment Report Institute of Geological Sciences*, No. 25.

PRUDDEN, H C, and EDWARDS, R A. 1994. Field excursion to the Minehead area, 5th January 1994. *Proceedings of the Ussher Society*, Vol. 8, 336–337.

RAMSBOTTOM, W H C, CALVER, M A, EAGAR, R M C, HODSON, F, HOLLIDAY, D W, STUBBLEFIELD, C J, and WILSON, R B. 1978. A correlation of Silesian rocks in the British Isles. *Special Report of the Geological Society of London*, No. 10.

RETALLACK, G J. 1988. Field recognition of palaeosols. 1–20 *in* Paleosols and weathering through geologic time. REINHARDT, J, and SIGLEO, W R (editors). *Special Paper of the Geological Society of America*, No. 216.

RICHARDSON, L. 1911. The Rhaetic and contiguous deposits of west, mid and part of east Somerset. *Quarterly Journal of the Geological Society of London*, Vol. 61, 385–424.

RICHARDSON, L. 1928. Wells and springs of Somerset. *Memoir of the Geological Survey of Great Britain.*

RIDING, J B. 1994. A palynological examination of four samples from the Bristol Channel. *British Geological Survey Technical Report*, WH/94/81R.

RILEY, N J. 1995. Foraminifera and algae from limestone clasts within the Permo-Trias of SW England. *British Geological Survey Technical Report*, WH/95/140R.

RØE, S L. 1987. Cross-strata and bedforms of probable transitional dune to upper-stage plane-bed origin from a Late Precambrian fluvial sandstone, northern Norway. *Sedimentology*, Vol. 34, 89–101.

SANDERSON, D J, and DEARMAN, W R. 1973. Structural zones of the Variscan fold belt in SW England, their location and development. *Journal of the Geological Society of London*, Vol. 129, 527–536.

SAUNDERSON, H C, and LOCKETT, F P J. 1983. Flume experiments on bedforms and structures at the dune–plane bed transition. 49–58 *in* Modern and ancient fluvial systems. COLLINSON, J D, and LEWIN, J (editors). *Special Publication of the International Association of Sedimentologists* No. 6.

SCRIVENER, R C, SHEPHERD, T J, and WARRINGTON, G. 1993. Genetic relationship of base and precious metal mineralization to Permo-Triassic basins in England. Geofluids '93, Extended Abstracts volume, 363–367.

SCRIVENER, R C, DARBYSHIRE, D P F, and SHEPHERD, T J. 1994. Timing and significance of crosscourse mineralisation in SW England. *Journal of the Geological Society of London*, Vol. 151, 587–590.

SELLWOOD, B W, DURKIN, M K, and KENNEDY, W J. 1970. Report of a field meeting on the Jurassic and Cretaceous rocks of Wessex. *Proceedings of the Geologists' Association*, Vol. 81, 715–732.

SHEARMAN, D J. 1967. On Tertiary fault movements in north Devonshire. *Proceedings of the Geologist's Association*, Vol. 78, 555–566.

SHENNAN, I. 1992. Late Quaternary sea-level changes and crustal movements in eastern England and eastern Scotland; and assessment of models of coastal evolution. *Quaternary International*, Vol. 15, 161–173.

SHENNAN, I, and WOODWORTH, P L. 1992. A comparison of late Holocene and twentieth-century sea-level trends from the UK and North Sea region. *Geophysical Journal International*, Vol. 109, 96–105.

SHERLOCK, R L, and HOLLINGWORTH, S E. 1938. Gypsum and anhydrite (3rd edition). *Special Report on the Mineral Resources of Great Britain, Memoir of the Geological Survey of Great Britain*, Vol. 3.

SHERRELL, F W. 1970.    Some aspects of the Triassic aquifer in east Devon and west Somerset.    *Quarterly Journal of Engineering Geology*, Vol. 2, 255–286.

SIMPSON, S. 1957.    On the trace fossil Chondrites.    *Quarterly Journal of the Geological Society of London*, Vol. 112, 475–500.

SIMPSON, S. 1959.    *Lexique Stratigraphique International. Fascicule 3aVI. Devonian.*    (Paris: Centre National de la Recherche Scientifique.)

SIMPSON, S. 1964.    The Lynton Beds of north Devon.    *Proceedings of the Ussher Society*, Vol. 1, 121–122.

SIMPSON, S. 1969.    Geology.    5–26 in *Exeter and its region.* BARLOW, F (editor). (Exeter: Exeter University Press.)

SIMPSON, S. 1971.    The Variscan structure of N. Devon.    *Proceedings of the Ussher Society*, Vol. 2, 249–252.

SMITH, N J P. 1993.    The case for exploration for deep plays in the Variscan fold belt and its foreland.    667–675 in *Petroleum geology of Northwest Europe: Proceedings of the 4th Conference.* PARKER, J R (editor). (London: Geological Society.)

SMITH, S A. 1990.    The sedimentology and accretionary styles of an ancient gravel-bed stream: the Budleigh Salterton Pebble Beds (Lower Triassic), southwest England.    *Sedimentary Geology*, Vol. 67, 199–219.

SMITH, S A, and EDWARDS, R A. 1991.    Regional sedimentological variations in Lower Triassic fluvial conglomerates (Budleigh Salterton Pebble Beds), southwest England: some implications for palaeogeography and basin evolution.    *Geological Journal*, Vol. 26, 65–83.

SOLLAS, W J. 1883.    The estuary of the Severn and its tributaries; an enquiry into the nature and origin of the tidal sediment and alluvial flats.    *Quarterly Journal of the Geological Society of London*, Vol. 39, 611–626.

STRONG, G E. 1995.    Petrography of Permo-Triassic and Devonian rock specimens from the Minehead area, Somerset.    *British Geological Survey Technical Report*, WG/95/10.

TALBOT, M R, HOLM, K, and WILLIAMS, M A J. 1994.    Sedimentation in low-gradient desert margin systems: a comparison of the Late Triassic of northwest Somerset (England) and the late Quaternary of east-central Australia.    97–117 in Palaeoclimate and basin evolution of playa systems. ROSEN, M R (editor). *Special Paper of the Geological Society of America*, No. 289.

TAPPIN, D R, CHADWICK, R A, JACKSON, A A, WINGFIELD, R T R, and SMITH, N J P. 1994.    *United Kingdom offshore regional report: the geology of Cardigan Bay and the Bristol Channel.*    (London: HMSO for the British Geological Survey.)

TAPPIN, D R, and DOWNIE, C. 1978.    New Tremadoc strata at outcrop in the Bristol Channel.    *Journal of the Geological Society of London*, Vol. 135, 321.

TAYLOR, S R. 1983.    A stable isotope study of the Mercia Mudstones (Keuper Marl) and associated sulphate horizons in the English Midlands.    *Sedimentology*, Vol. 30, 11–31.

THOMAS, A N. 1940.    The Triassic rocks of north-west Somerset.    *Proceedings of the Geologists' Association*, Vol. 51, 1–43.

THOMAS, J B, MARSHALL, J, MANN, A L, SUMMONS, R E, and MAXWELL, J R. 1993.    Dinosteranes (4, 23, 24-trimethylsteranes) and other biological markers in dinoflagellate-rich marine sediments of Rhaetian age.    *Organic Geochemistry*, Vol. 20, 91–104.

THOMAS, J M. 1963a.    The Culm Measures succession in north-east Devon and north-west Somerset.    *Proceedings of the Ussher Society*, Vol. 1, 63–64.

THOMAS, J M. 1963b.    Sedimentation in the Lower Culm Measures around Westleigh, North-east Devon.    *Proceedings of the Ussher Society*, Vol. 1, 71–72.

THOMSON, G F, CORNWELL, J D, and COLLINSON, D W. 1991.    Magnetic characteristics of some pyrrhotite-bearing rocks in the United Kingdom.    *Geoexploration*, Vol. 27, 23–41.

TUNBRIDGE, I P. 1978.    North Devon: Lower and Middle Devonian. Lynton Beds and Hangman Sandstone Group.    8–13 in *International Symposium on the Devonian System. A field guide to selected areas of South-West England.* SCRUTTON, C T (editor). (Palaeontological Association.)

TUNBRIDGE, I P. 1981.    Sandy high-energy flood sedimentation — some criteria for recognition, with an example from the Devonian of S.W. England.    *Sedimentary Geology*, Vol. 28, 79–95.

TUNBRIDGE, I P. 1984.    Facies model for a sandy ephemeral stream and clay playa complex; the Middle Devonian Trentishoe Formation of north Devon, U.K.    *Sedimentology*, Vol. 31, 697–715.

TUNBRIDGE, I P. 1986.    Mid-Devonian tectonics and sedimentation in the Bristol Channel area.    *Journal of the Geological Society of London*, Vol. 143, 107–115.

TUWENI, A O, and TYSON, R V. 1994.    Organic facies variations in the Westbury Formation (Rhaetic, Bristol Channel, SW England).    *Organic Geochemistry*, Vol. 21, 1001–1014.

USSHER, W A E. 1875.    On the subdivisions of the Triassic rocks between the coast of west Somerset and the south coast of Devon.    *Geological Magazine*, Vol. 12, 163–168.

USSHER, W A E. 1876.    On the Triassic rocks of Somerset and Devon.    *Quarterly Journal of the Geological Society of London*, Vol. 32, 367–394.

USSHER, W A E. 1877.    A classification of the Triassic rocks of Devon and west Somerset.    *Transactions of the Devonshire Association for the Advancement of Science*, Vol. 9, 392–399.

USSHER, W A E. 1878.    Chronological value of the Triassic strata of the south-west counties.    *Quarterly Journal of the Geological Society of London*, Vol. 34, 459–470.

USSHER, W A E. 1889.    The Triassic rocks of west Somerset and the Devonian rocks on their borders.    *Proceedings of the Somersetshire Archaeological and Natural History Society*, Vol. 15, 1–36.

WARRINGTON, G. 1974.    Studies in the palynological biostratigraphy of the British Trias. I. Reference sections in west Lancashire and north Somerset.    *Review of Palaeobotany and Palynology*, Vol. 17, 133–147.

WARRINGTON, G. 1981.    The indigenous micropalaeontology of British Triassic shelf sea deposits.    61–70 in *Microfossils from Recent and fossil shelf sea areas.* NEALE, J W, and BRASIER, M D (editors). (Chichester: Ellis Horwood.)

WARRINGTON, G. 1983.    Appendix 4. Mesozoic micro-palaeontological studies.    131–132 in Geology of the country around Weston-super-Mare. WHITTAKER, A, and GREEN, G W. *Memoir of the Geological Survey of Great Britain*, Sheet 279, with parts of sheets 263 and 295 (England and Wales).

WARRINGTON, G. 1985.    Appendix 1. Palynology of the Permo-Triassic and lower Jurassic succession.    84–85 in Geology of the country around Taunton and the Quantock Hills. EDMONDS, E A, and WILLIAMS, B J. *Memoir of the Geological Survey of Great Britain*, Sheet 295 (England and Wales).

WARRINGTON, G. 1994. Palynology report: Mercia Mudstone Group (Triassic), between Porlock and Dunster, west Somerset (Sheet 278: Minehead). *British Geological Survey Technical Report*, WH/94/293R.

WARRINGTON, G, and IVIMEY-COOK, H C. 1995. The Late Triassic and Early Jurassic of coastal sections in west Somerset and South and Mid-Glamorgan. 9–30 in *Field geology of the British Jurassic*. TAYLOR, P D (editor). (London: Geological Society.)

WARRINGTON, G, and OWENS, B (compilers). 1977. Micro-palaeontological biostratigraphy of offshore samples from south-west Britain. *Report of the Institute of Geological Sciences*, No. 77/7.

WARRINGTON, G, and WHITTAKER, A. 1984. The Blue Anchor Formation (late Triassic) in Somerset. *Proceedings of the Ussher Society*, Vol. 6, 100–107.

WARRINGTON, G, AUDLEY-CHARLES, M G, ELLIOTT, R E, EVANS, W B, IVIMEY-COOK, H C, KENT, P E, ROBINSON, P L, SHOTTON, F W, and TAYLOR, F M. 1980. A correlation of Triassic rocks in the British Isles. *Special Report of the Geological Society of London*, No.13.

WARRINGTON, G, COPE, J C W, and IVIMEY-COOK, H C. 1994. St Audrie's Bay, Somerset, England: a candidate Global Stratotype Section and Point for the base of the Jurassic System. *Geological Magazine*, Vol. 131, 191–200.

WARRINGTON, G, IVIMEY-COOK, H C, EDWARDS, R A, and WHITTAKER, A. 1995. The late Triassic–early Jurassic succession at Selworthy, west Somerset, England. *Proceedings of the Ussher Society*, Vol. 8, 426–432.

WATERS, R A, and LAWRENCE, D J D. 1987. Geology of the South Wales Coalfield, Part III, the country around Cardiff (3rd edition). *Memoir of the British Geological Survey*, Sheet 263 (England and Wales).

WEBBY, B D. 1965. The stratigraphy and structure of the Devonian rocks in the Brendon Hills, west Somerset. *Proceedings of the Geologists' Association*, Vol. 76, 39–60.

WEDLAKE, A L. 1950. Mammoth remains and Pleistocene implements found on the west Somerset coast. *Proceedings of the Somerset Archaeological Society*, Vol. 95, 167–168.

WEDLAKE, A L, and WEDLAKE, D J. 1963. Some palaeoliths from the Doniford Gravels on the coast of west Somerset. *Proceedings of the Somerset Archaeological and Natural History Society*, Vol. 107, 93–100.

WEEDON, G P. 1986. Hemipelagic shelf sedimentation and climatic cycles: the basal Jurassic (Blue Lias) of South Britain. *Earth and Planetary Science Letters*, Vol. 76, 321–335.

WENTWORTH, C K. 1922. A scale of grade and class terms for clastic sediments. *Journal of Geology*, Vol. 30, 377–392.

WHITTAKER, A. 1972. The Watchet Fault — A post-Liassic transcurrent reverse fault. *Bulletin of the Geological Survey of Great Britain*, No. 41, 75–80.

WHITTAKER, A. 1976. Notes on the Lias outlier near Selworthy, west Somerset. *Proceedings of the Ussher Society*, Vol. 3, 355–359.

WHITTAKER, A. 1978a. Discussion of the gravity gradient across Exmoor, north Devon. *Journal of the Geological Society of London*, Vol. 135, 353–354.

WHITTAKER, A. 1978b. The lithostratigraphical correlation of the uppermost Rhaetic and lowermost Liassic strata of the west Somerset and Glamorgan areas. *Geological Magazine*, Vol. 115, 63–67.

WHITTAKER, A, and GREEN, G W. 1983. Geology of the country around Weston-super-Mare. *Memoir of the Geological Survey of Great Britain*, Sheet 279, with parts of sheets 263 and 295 (England and Wales).

WHITTAKER, A, and SCRIVENER, R C. 1982. The Knap Farm Borehole at Cannington Park, Somerset. *Report of the British Geological Survey*, No. 82/5.

WILKINSON, I P. 1994. Jurassic microfaunas from a suite of samples off Minehead. *British Geological Survey Technical Report*, WH/94/65R.

WILKINSON, I P, and HALLIWELL, G P (compilers). 1980. Offshore micropalaeontological biostratigraphy of southern and western Britain. *Report of the Institute of Geological Sciences*, No. 79/9.

WILLIAMS, A T, and DAVIES, P. 1990. A coastal hard rock sediment budget for the inner Bristol Channel. 474–479 in *Sediment transport modelling*. WANG, S (editor). (American Society of Civil Engineers.)

WILSON, H. 1995. The coastal geomorphology of Exmoor. 26–32 in *The changing face of Exmoor*. BINDING, H (editor). (Tiverton: Exmoor Books).

WOOLLAM, R, and RIDING, J B. 1983. Dinoflagellate cyst zonation of the English Jurassic. *Report of the Institute of Geological Sciences*, No. 83/2.

WOODWARD, H B. 1893. The Lias of England and Wales (Yorkshire excepted). *Memoir of the Geological Survey of Great Britain*.

WRIGHT, V P, and SANDLER, A. 1994. A hydrogeological model for the early diagenesis of Late Triassic alluvial sediments. *Journal of the Geological Society of London*, Vol. 151, 897–200.

WRIGHT, V P, and TUCKER, M E. 1991. Calcretes: an introduction. 1–22 in *Calcretes. Reprint Series Volume 2 of the International Association of Sedimentologists*. WRIGHT, V P, and TUCKER, M E (editors). (Oxford: Blackwells.)

# AUTHOR CITATIONS FOR FOSSIL SPECIES

To satisfy the rules and recommendations of the international codes of botanical and zoological nomenclature, authors of cited species are listed below.

## Chapter 4 Devonian

*Chonetes sarcinulatus* (Schlotheim, 1820)
*Platyorthis longisulcata* (Phillips, 1841)
*Pseudosporochnus nodosus* Leclercq & Banks, 1962
*Subcuspidella lateincisa* (Scupin, 1900)

## Chapter 5 Permo-Triassic

*Acteonina fusiformis* (Moore, 1861)
*Acteonina ovalis* (Moore, 1861)
*Acteonina oviformis* (Moore, 1861)
*Annulithus arkelli* Rood, Hay & Barnard, 1973
*Atreta intusstriata* (Emmrich, 1853)
*?Beaumontella caminuspina* (Wall) Below, 1987
*Beaumontella langii* (Wall) Below, 1987
*Calamospora mesozoica* Couper, 1958
*Ceratodus latissimus* Agassiz, 1838
*Chasmatosporites magnolioides* (Erdtman) Nilsson, 1958
*Chlamys valoniensis* (Defrance, 1825)
*Cingulizonates rhaeticus* (Reinhardt) Schulz, 1967
*Classopollis torosus* (Reissinger) Balme, 1957
*Contignisporites problematicus* (Couper) Döring, 1965
*Converrucosisporites luebbenensis* Schulz, 1967
*Cornutisporites seebergensis* Schulz, 1962
*Cymatiosphaera polypartita* Morbey, 1975
*Dacryomya titei* (Moore, 1861)

*Dapcodinium priscum* Evitt, 1961 emend. Below, 1987
*Eotrapezium concentricum* (Moore, 1861)
*Eucommiidites major* Schulz, 1967
*Geopollis zwolinskae* (Lund) Brenner, 1986
*'Gervillia' praecursor* Quenstedt, 1858
*Gliscopollis meyeriana* (Klaus) Venkatachala, 1966
*Granuloperculatipollis rudis* Venkatachala & Góczán emend. Morbey, 1975
*Gyrolepis alberti* Agassiz, 1835
*Hybodus minor* Agassiz, 1843
*Kraeuselisporites reissingeri* (Harris) Morbey, 1975
*Leptolepidites argenteaeformis* (Bolchovitina) Morbey, 1975
*Limbosporites lundbladii* Nilsson, 1958
*Liostrea hisingeri* (Nilsson, 1831)
*Lissodus minimus* (Agassiz, 1839)
*Lunatisporites rhaeticus* (Schulz) Warrington, 1974
*Lycopodiacidites rhaeticus* Schulz, 1967
*Lyriomyophoria postera* (Quenstedt, 1856)
*Microreticulatisporites fuscus* (Nilsson) Morbey, 1975
*'Modiolus' sodburiensis* Vaughan, 1904
*'Natica' oppelii* Moore, 1861
*Nevesisporites bigranulatus* (Levet-Carette) Morbey, 1975
*Ovalipollis pseudoalatus* (Thiergart) Schuurman, 1976
*Pinuspollenites minimus* (Couper) Kemp, 1970
*Placunopsis alpina* (Winkler, 1859)
*'Pleurophorus' elongatus* Moore, 1861
*Porcellispora longdonensis* (Clarke) Scheuring emend. Morbey, 1975
*Protocardia rhaetica* (Merian, 1853)
*Protohaploxypinus microcorpus* (Schaarschmidt) Clarke, 1965
*Pteromya crowcombeia* Moore, 1861
*Quadraeculina anellaeformis* Maljavkina, 1949
*Rhaetavicula contorta* (Portlock, 1843)
*Rhaetipollis germanicus* Schulz, 1967
*Rhaetogonyaulax rhaetica* (Sarjeant) Loeblich & Loeblich, 1968 emend. Below, 1987
*Ricciisporites tuberculatus* Lundblad, 1964

*Schizosphaerella punctulata* Deflandre & Dangeard, 1938
*Sphaerinvia piai* Vachard, 1980
*Tasmanites newtoni* Wall, 1965
*Todisporites minor* Couper, 1958
*Tsugaepollenites ?pseudomassulae* (Mädler) Morbey, 1975
*Tutcheria cloacina* (Quenstedt, 1856)
*Vesicaspora fuscus* (Pautsch) Morbey, 1975
*Vitreisporites pallidus* (Reissinger) Nilsson, 1958

## Chapter 6 Jurassic

*Beaumontella langii* (Wall) Below, 1987
*Calcirhynchia calcaria* S S Buckman, 1918
*Caloceras intermedium* (Portlock, 1843)
*Classopollis torosus* (Reissinger) Balme, 1957
*Caloceras johnstoni* (J de C Sowerby, 1824)
*Coroniceras hyatti* Donovan, 1952
*Coroniceras rotiforme* (J de C Sowerby, 1824)
*Gliscopollis meyeriana* (Klaus) Venkatachala, 1966
*Gryphaea arcuata* Lamarck, 1801
*Kraeuselisporites reissingeri* (Harris) Morbey, 1975
*Leptolepidites argenteaeformis* (Bolchovitina) Morbey, 1975
*Kraeuselisporites reissingeri* (Harris) Morbey, 1975
*Liostrea hisingeri* (Nilsson, 1831)
*Modiolus minimus* J Sowerby, 1818
*Nevesisporites bigranulatus* (Levet-Carette) Morbey, 1975
*Plagiostoma giganteum* J Sowerby, 1814
*Psiloceras planorbis* (J de C Sowerby, 1824)
*Pteromya tatei* (Richardson & Tutcher, 1914)
*Quadraeculina anellaeformis* Maljavkina, 1949
*Schlotheimia extranodosa* (Waehner, 1886)
*Schlotheimia striatissima* (Quenstedt, 1883)

## Chapter 8 Quaternary

*Elephas primigenius* Blumenbach, 1799
*Scrobicularia piperata* (Gmelin, 1788)

# INDEX

**BRITISH GEOLOGICAL SURVEY**

Keyworth, Nottingham NG12 5GG
0115 936 3100

Murchison House, West Mains Road, Edinburgh EH9 3LA
0131 667 1000

London Information Office, Natural History Museum
Earth Galleries, Exhibition Road, London SW7 2DE
020 7589 4090

The full range of Survey publications is available through the
Sales Desks at Keyworth and at Murchison House, Edinburgh,
and in the BGS London Information Office in the Natural
History Museum (Earth Galleries). The adjacent bookshop
stocks the more popular books for sale over the counter. Most
BGS books and reports can be bought from The Stationery
Office and through Stationery Office agents and retailers.
Maps are listed in the BGS Map Catalogue, and can be bought
together with books and reports through BGS-approved
stockists and agents as well as direct from BGS.

*The British Geological Survey carries out the geological survey of Great
Britain and Northern Ireland (the latter as an agency service for the
government of Northern Ireland), and of the surrounding continental
shelf, as well as its basic research projects. It also undertakes
programmes of British technical aid in geology in developing countries
as arranged by the Department for International Development and
other agencies.*

*The British Geological Survey is a component body of the Natural
Environment Research Council.*

Published by The Stationery Office and available from:

**The Publications Centre**
(mail, telephone and fax orders only)
PO Box 276, London SW8 5DT
Telephone orders/General enquiries 0870 600 5522
Fax orders 0870 600 5533

www.tso-online.co.uk

**The Stationery Office Bookshops**
123 Kingsway, London WC2B 6PQ
020 7242 6393  Fax 020 7242 6412
68–69 Bull Street, Birmingham B4 6AD
0121 236 9696  Fax 0121 236 9699
33 Wine Street, Bristol BS1 2BQ
0117 926 4306  Fax 0117 929 4515
9–21 Princess Street, Manchester M60 8AS
0161 834 7201  Fax 0161 833 0634
16 Arthur Street, Belfast BT1 4GD
028 9023 8451  Fax 028 9023 5401
The Stationery Office Oriel Bookshop
18–19 High Street, Cardiff CF1 2BZ
029 2039 5548  Fax 029 2038 4347
71 Lothian Road, Edinburgh EH3 9AZ
0870 606 5566  Fax 0870 606 5588

**The Stationery Office's Accredited Agents**
(see Yellow Pages)

*and through good booksellers*